RESOURCE MANAGEMENT, URBANIZATION AND GOVERNANCE IN HONG KONG AND THE ZHUJIANG DELTA

Resource Management, Urbanization and Governance in Hong Kong and the Zhujiang Delta

Edited by

Kwan-yiu Wong and Jianfa Shen

The Chinese University Press

Resource Management, Urbanization and
Governance in Hong Kong and the Zhujiang Delta
Edited by Kwan-yiu Wong and Jianfa Shen

© **The Chinese University of Hong Kong** 2002

ISBN 962–996–026–5

THE CHINESE UNIVERSITY PRESS
The Chinese University of Hong Kong
SHA TIN, N.T., HONG KONG
Fax: +852 2603 6692
    +852 2603 7355
E-mail: cup@cuhk.edu.hk
Web-site: www.chineseupress.com

Printed in Hong Kong

Contents

Preface

The Zhujiang Delta region has experienced astonishing growth since the late 1970s. In many ways, such development is closely related with that of Hong Kong. The Commission on Strategic Development of the Hong Kong Special Administrative Region (HKSAR) recognized in its recent report "Bringing the Vision to Life: Hong Kong's Long-Term Development Needs and Goals" that capitalizing on the linkage with mainland China, and with the Zhujiang Delta (Pearl River Delta) region in particular, is one of the key strategies for the future long-term development of the HKSAR.

The pace of development in the Zhujiang Delta has exceeded the expectations of the outside world and even the planners. The increasing interaction and integration with Hong Kong are equally obvious. These processes have continued after the return of sovereignty of Hong Kong to China on July 1, 1997. But there are emerging issues and concerns of effective governance and sustainability that will affect the further development and prosperity of the region. Thus, there is an urgent need to monitor and analyze its development. Flexible models and forms of governance, integration and coordination will be required.

The research project "Urbanization and Governance in the Zhujiang Delta: Analysis and Policy Implications," with funding support from the Research Grant Council of Hong Kong (RGC reference no. CUHK4017/ 98H), was conducted by investigators Kwan-yiu Wong, David K Y Chu and Jianfa Shen of the Department of Geography, The Chinese University of Hong Kong during the 1998–2002 period. The research used both modelling and qualitative approaches to analyze the process of development and urbanization in the region and to develop policy recommendations on urban/regional governance in this rapidly growing region. Some of the research findings from this project have been published in eight academic journals such as *The Geographical Journal* ("The spatial dynamics of foreign investment in the Pearl River Delta, south China," 2000, 166(4), 312–22) and *Environment and Planning A* ("Regional polarization under the socialist-market system since 1978 — a case study of Guangdong province in south China," 2001, 33(1), 97–119). Several scholars from Hong Kong and mainland China have directly participated in the project and have made valuable contributions. Particular thanks are due to Prof. Chaolin Gu, Dr. Zhiqiang Feng, Dr. Chaoyong Huang, Dr. Feng Zhen, Miss Joanna Sin, Mr. Tianxin Hu and Ms. Chunxiao Huang.

The current volume is a collective effort by scholars from Hong Kong, mainland China, the United States and Germany, as well as the investigators in the above RGC project who presented their recent research findings at the International Workshop on Resource Management, Urbanization and Governance in Hong Kong and the Zhujiang Delta on 23–24 May 2000. The fourteen chapters of this book cover economic integration, urbanization and regional development, resource management and urban/regional governance in the Hong Kong–Zhujiang Delta region. The workshop was jointly organized by United College and the Department of Geography, The Chinese University of Hong Kong. We are grateful to our officiating guests, Dr. Edgar W K Cheng, Professor Kenneth Young and Mr. C S Shum, keynote speaker Professor Yun-wing Sung, session chairmen Professor Yee Leung, Professor Roger Chan, Professor Kin-che Lam, Professor Anthony Yeh and Prof. Alvin So, as well as to other speakers and participants at the workshop for their support. Thanks are also due to the contributing authors to this volume for carefully preparing the final versions of their papers by incorporating valuable comments from the workshop participants. Funding support from RGC grant CUHK4017/98H and the Endowment Fund of United College is also gratefully acknowledged.

Kwan-yiu Wong and Jianfa Shen
April 2002

1

Opening Addresses

Address by Dr. Edgar W. K. Cheng,
Head, Central Policy Unit, HKSAR Government

The Hong Kong/Pearl River Delta (HK/PRD) city-region is perhaps the only one of its kind in terms of the excitement and interest that it can generate simultaneously among three types of people — businessmen, government officials and academic researchers.

As an economic entity, the HK/PRD region is composed of 35 million people, with an estimated combined gross output of US$244 billion in 1999. Most parts grew at a dizzying pace of 13.9% per year in real terms during the 1995–1998 period. It is a growth market with the potential for immense productivity gains, not a mature or replacement market with more or less predictable and average bottom lines. For many, the place may appear to be quite ordinary, but in terms of business opportunities, it is simply extraordinary.

As a political entity, it is one ultimate sovereignty dichotomized into two systems with vastly different theories and practices of governance. One system, that of Hong Kong, is endowed with great economic sophistication, but this can be attributed to the other system that acts as its guarantor of socio-political continuity under "One Country Two Systems." This is unlike most other geo-economic contexts in which preponderant political power is vested where economic might is greater, and city governments generally make their influence felt in nearby regions or the hinterland. However, this is not the case in the HK/PRD setting. HK government officials do not hold sway over their regional counterparts and thus they must treat them duly as important equals. Likewise, PRD officials cannot leverage any of their political prerogatives on HK, if only because of the sheer weight of the economic power vested in the Hong Kong Special Administrative Region. This interesting balance makes cooperation between HK and the rest of the PRD region at the government level possible on the one hand and most delicate on the other. It requires great skills, patience, mutual understanding and respect. It is a challenge to all officials in the HK/PRD region who look beyond the confines of their common political-administrative boundaries to bring synergies and mutual benefits to the entire city-region. Moreover, a smooth working relationship between officials of each side presupposes a deep knowledge of the realities of the other side. Learning is imperative.

For scholars and researchers, the great appeal of the HK/PRD region lies in the intriguing, stark juxtaposition of similarities and differences between HK and the PRD. Nowhere in the world can one find two contiguous areas with so many economic, cultural, family and emotional ties and also such deep disparities in political and social philosophies and outlook. But all these differences and similarities are becoming blurred in the context of rapid IT advancements and globalization. What are the trends, consistencies and laws governing these changes? By their very nature scholars and researchers find these questions most intellectually stimulating.

Hence the interest shown by people of different backgrounds to attend and share insights at this highly noteworthy conference. It is very heartening for me to see such interest among different stakeholders represented at today's international workshop on HK-PRD regional development.

The history of development is replete with horrible mistakes, mistakes that have been extremely costly both in material and human terms. One can recall the environmental destruction or the social decay accompanying much of the world's economic development in the last fifty years, both in advanced and developing countries. It would be simplistic, yet not entirely wrong, to say that the agents of the forces of development were reckless, and those who were supposed to provide enlightenment were ineffectual. Now, we the stakeholders — people who inhabit the HK/PRD region — do not want similar mistakes to be repeated as we undertake the development of the vast potentials of this city-region. For this, we must deepen and broaden our knowledge of the area through rigorous research and debate, in order to improve the quality of our policies.

It thus behooves us, as scholars, researchers, government and private sector decision-makers, to work hand in hand to insure that we understand well the nature of the development problems at hand, and the appropriateness and limitations of the solutions we propose to bring to bear on these problems, so as to derive mutually beneficial outcomes.

This region is our region. It belongs to all the people in the HK/PRD area. This is where we live and die, prosper or decline, shine or fall behind. So much is at stake. The implications are clear.

I congratulate the successful convening of this workshop, and eagerly await the flow of constructive ideas. Thank you.

Address by Professor Kenneth Young, Pro-Vice-Chancellor of The Chinese University of Hong Kong

I am delighted to welcome you all to the International Workshop on Resource Management, Urbanization and Governance in Hong Kong and the Zhujiang Delta, jointly organized by United College and the Department of Geography of The Chinese University of Hong Kong.

The Zhujiang Delta region is one of the most rapidly developing areas of China, and its growth in many ways is closely integrated with the development of Hong Kong. As a financial and business centre in the Asia-Pacific region, Hong Kong is, and will continue to be, a hub and focus of the emerging Zhujiang Delta metropolitan area. In particular, it is well accepted that Hong Kong will take on roles in capital, technology, management and informatics. On the other hand, developments in the Zhujiang Delta will affect Hong Kong's development strategies in the years ahead.

Thus there are opportunities for better and closer coordination between Hong Kong and the Zhujiang Delta to achieve more concerted development of the region. A key issue is whether such integration will enhance the comparative advantage of this consolidated economic entity in the global context. And I want to emphasize the increasingly recognized need to view this region from a global perspective.

Effective governance and social sustainability are key concerns of governments in urban and regional development. Commentators have noted that the Zhujiang Delta lacks an effective and flexible form of governance that is capable of coordinating, for example, different levels of government, local and foreign businesses, and research and planning institutions. In part, the problems arise because of the boundary that separates the two systems within one country, the two complementary systems organized according to rather different principles. This has led to many serious problems. The more obvious include, for example, environmental pollution and degradation, the over-development of land resources (that is, land levelled but not put to good use), a redundancy of investment, overlapping responsibilities and competing domains with little regard to economic costs (as in the case of infrastructure development).

This Workshop will address these issues, in order to investigate the roots of the problems and to look into possible ways to ameliorate and rectify them. We welcome the experienced and distinguished scholars and government officials gathered together here to share their insights. Much will come out of this Workshop, not only in terms of scholarship, but more importantly in terms of advice to the authorities. It is for this reason that we are especially grateful to Dr. Cheng, Head of the Central Policy Unit, for being with us today, and for delivering the opening address.

I wish the Workshop every success, and I am sure that the region's development will benefit from the deliberations.

Address by Mr. Choi-sang Shum, Chairman of the Board of Trustees of United College, The Chinese University of Hong Kong

I am deeply honoured to join all of you at the Opening Ceremony of this very important and timely Workshop.

I believe none of you will dispute the importance and timeliness of the Workshop. It will help us to look into our future at this very crucial time during the strategic development of Hong Kong and the Pearl River Delta. I believe that following the closer integration of the two areas it is a future full of promise for the people of both Hong Kong and South China. I find it most encouraging to see United College and the Department of Geography of The Chinese University, and all of Hong Kong, address the various issues involved and respond to the challenges and opportunities arising therefrom.

I believe the Workshop reflects a wider vision of life in both Hong Kong and the Pearl River Delta, not just concerned with our economic interests, but also with improving the quality of life — of improving our environment, of quality development and governance, and of awakening to our own cultural identity and mission. The Board of Trustees of United

College is indeed proud to be able to identify itself with the aims of this Workshop through the sponsorship of its Endowment Fund.

On this special occasion, I would like to congratulate all those who have helped organize the Workshop and those who are participating in it, and I wish the Workshop every success.

Thank you very much.

PART I

Keynote Speech

Economic Integration of Hong Kong and the Zhujiang Delta

Yun-wing Sung

Department of Economics, The Chinese University of Hong Kong

Introduction

Though the Zhujiang Delta is the natural hinterland of Hong Kong, the framework of "one country, two systems" has imposed many restrictions on the form of economic integration between Hong Kong and its hinterland. Despite the reversion of Hong Kong to China in mid-1997, for practical purposes Hong Kong is still an independent economic entity with its own currency, customs and immigration controls.

There are many barriers to the economic integration of Hong Kong and the Zhujiang Delta, some of which will persist until 2047 or even beyond. For instance, it is virtually impossible for Hong Kong and the Zhujiang Delta (or the mainland as a whole) to form a formal trade bloc on the model of the EU (European Union). Even the establishment of a border industrial park is fraught with difficulties.

Though "old-style" integration based on discriminatory preferences on trade in goods is impractical, the reversion of Hong Kong presents opportunities for a "new style" integration through policy co-ordination. The reversion of Hong Kong, the AFC (Asian Financial Crisis), and China's impending entry into the WTO have promoted such co-ordination.

Of late, there has been much discussion among APEC (Asian Pacific Economic Cooperation) members about forming "new style" FTAs (Free Trade Areas) that concentrate on trade in services rather than trade in goods.

For example, Japan and Singapore are considering such a FTA. The possibility of a "new style" FTA for Hong Kong and the mainland will be discussed below.

This article is organized as follows. After the introductory section, section two will discuss the limits to integration imposed by the framework of "one country, two systems". Section three will discuss opportunities for policy co-ordination. Section four will discuss the possibilities for a "new style" FTA, and section five will present some conclusions.

Limits to economic integration

Hong Kong's reversion does not change the framework of Hong Kong-mainland economic relations, which was laid down in the Sino-British Declaration of 1984 and the Basic Law. For all practical purposes, Hong Kong is a separate economic entity. It is a separate customs territory, with its own border and immigration controls. It has its own currency which is linked to the US dollar and it runs its own monetary and fiscal policies. It does not pay any taxes to the central government.

Despite Hong Kong's reversion to China, there is very little hope that Hong Kong and the mainland can form some kind of trade bloc, such as a Customs Union or a Common Market. A Customs Union is out of the question as it requires that Hong Kong and the mainland levy the same external tariffs, which in Hong Kong is zero as it is a free port. Both the 1984 Sino-British Declaration and the Basic Law have specified that Hong Kong will be a free port. A Common Market is even more problematic as it requires free migration within the bloc and this runs counter to Hong Kong's very strict immigration controls against the mainland. Even a traditional FTA, which has the lowest degree of formal economic integration, is impractical because the mainland would lose tariff revenue and there would be no offsetting gains in mainland exports as Hong Kong is already a free port. A FTA on trade in goods is not a "win-win" proposition. A "new-style" FTA on investment and services trade is more promising, as will be discussed later.

A south China trading bloc?

A Hong Kong-Guangdong trading bloc is even more problematic as it would imply the building of a fence to separate Guangdong from the rest of China.

Otherwise, Guangdong could not have different tariffs from the rest of the mainland.

Even the formation of a Customs Union involving Hong Kong and Shenzhen is unrealistic. Shenzhen has built a 'second line' separating itself from the rest of China and it aspires to become a free trade zone. Even if Shenzhen were to become a free trade zone, Hong Kong and Shenzhen would still be separate trade entities. Trade between Hong Kong and Shenzhen would be similar to trade between Hong Kong and any other free trade zone, for example, trade between Hong Kong and Singapore. Theoretically, Hong Kong and Shenzhen could enter the WTO as a single customs territory. The two regions then would have to bargain as a single entity in world trade and also to agree on the sharing of textile and clothing quotas. A prerequisite for that is that Beijing would have to relinquish control of Shenzhen's external economic affairs. China would also have to enter the WTO as the mainland less Shenzhen, because the bargaining position of Shenzhen in world trade may differ from that of Beijing.

Another problem is that the union would undermine Hong Kong's autonomy. Hong Kong's freedom in external economic affairs is guaranteed by both the Sino-British agreement and the Basic Law, and Hong Kong is a member of the WTO. A trading bloc of Shenzhen and Hong Kong would provide an institutional channel for Beijing to encroach on Hong Kong's autonomy. Hong Kong has the world's largest clothing quota which it does not want to share with anyone else.

Even if Shenzhen were to become a free trade zone, Hong Kong and Shenzhen would have to maintain their border controls against each other. To qualify for WTO membership, Hong Kong must be able to effectively distinguish between goods made in Hong Kong and those that are made in Shenzhen and elsewhere. The abolition of Hong Kong border controls to Shenzhen would jeopardize Hong Kong's WTO membership and Hong Kong's textile and clothing quotas. The abolition of Shenzhen's border controls against Hong Kong goods would also pose complications for China's WTO membership.

Moreover, Hong Kong and Shenzhen have different import prohibitions. For example, Shenzhen does not allow free importation of political or religious literature from Hong Kong. Hong Kong has to maintain controls on immigration from the mainland for reasons of overcrowding. The need for strict controls is recognised by both the Sino-British agreement and the Basic Law. Shenzhen also needs to maintain some controls over the entry of Hong Kong residents to prevent "undesirable" people from entering China. For example, during the riots in Shenzhen in August 1992, Shenzhen forbade the entry of Hong Kong reporters. Even if Shenzhen were to become a free trade zone, immigration and customs controls still would have to be enforced.

Bonded Areas

After Deng's 1992 southern tour, China established thirteen bonded areas along the coast; three of these areas are in the Pearl River Delta, namely, Guangzhou, and Shenzhen's Futian and Shatoujiao. Moreover, Shenzhen plans to 'open up the first line and tighten up the second line', thus transforming Shenzhen into a huge bonded area.

Besides bonded areas, there are also many bonded warehouses and bonded factories in the Pearl River Delta. Products imported into bonded areas, warehouses or factories are temporarily exempted from tariffs if they are later exported. However, if they are sold in the domestic market, then full tariffs are levied.

A bonded area faces international prices and is an area of free trade because no tariffs are levied. Bonded warehouses, factories or areas are good for developing processed exports because imports for processing are exempted from tariffs. Bonded areas are also good for developing the ancillary services of export processing, namely, storage, transportation, and packaging.

In non-Communist developing countries, tariffs are usually the main factor that distorts the functioning of the economy. Bonded areas would remove this distortion. In the traditional command economy of China, administrative intervention was the main distortion and tariffs were only of secondary importance. China's open policy thus started with the creation of special economic zones that were exempted from many of the controls of the plan. By the early 1990s, a nascent market economy had developed in the coastal open areas, and the establishment of bonded areas was the next logical step.

China's bonded areas are completely enclosed to prevent smuggling and to facilitate management. Workers pass through checkpoints to work inside the bonded areas. There are no residents inside these areas. This arrangement minimizes the need to import consumer durables for use within the area, thus minimizing the losses in tariff revenue.

The Futian Bonded Area

Among the many bonded areas of China, Shenzhen's Futian has a great potential because it is adjacent to Hong Kong. The Futian zone is connected to Hong Kong by the Lok Ma Chau bridge, thus there is no need to pass through Shenzhen customs. Futian is within an hour's drive from most parts of Hong Kong and foreigners can apply for a visa upon entry. It is intended to be a convenient place for contacts between Chinese and foreigners, and expected to develop into a services centre specializing in international trade, exhibitions, information, finance and insurance. The Weigaoqiao bonded area in Shanghai, which has its own port, has a similar goal.

However, the development of both Futian and Weigaoqiao has failed the expectations of the planners. This is largely attributable to a mistaken strategy. From international experience, while bonded areas are good for export processing and storage, they are not suited to trading, finance, and related services. Many Chinese planners have the mistaken notion that the tariff-free status of bonded areas is important for the development of entrepôt trade. However, due to the "hub-and-spoke" pattern of transportation, an entrepôt has to be a major transportation hub. Futian and the other bonded areas are not major transportation hubs because they are far away from the city centre so as to ensure that they can be easily fenced in.

As for trading, finance, and related services, where deals are negotiated and trade documents processed can be quite far away from where the goods are processed or stored. For instance, many deals concluded in Hong Kong involve trade that does not touch upon Hong Kong at all. The negotiation of deals and processing of trade documents usually take place in trading and business centres due to the economies of agglomeration of such activities. Bonded areas are good for storage or export processing rather than service activities.

Another problem with bonded areas in China is that local governments want to use them as an excuse to give special tax breaks to attract foreign investment. While special tax breaks had a role at the initial stage of the open policy, a mature investment environment should not rely on tax breaks. Moreover, China is preparing to enter the WTO which requires uniform or national treatment for all, and it will need to phase out the special tax breaks for the SEZs and foreign investors. However, competition for foreign investment among localities in China is very keen, and local authorities often establish bonded areas as a pretext to evade taxes or for outright smuggling. The result is that the China Customs Administration makes controls in bonded areas so tight and inflexible that investors have few incentives to operate there. For instance, Hong Kong investors complain that it is easier to enter Futian from Shenzhen rather than directly from Hong Kong despite the existence of a direct entrance.

Bonded areas are run by special Management Boards, and this means that foreign investors have to deal with an additional layer of bureaucracy. In the case of Futian, the Board appears to be more interested in short-term gains of its own than in long-run economic development. In fact, long before there were any plans for the Futian bonded area, Hong Kong investors had leased the land in the Futian area because of its development potential due to its proximity to Hong Kong. The action by Hong Kong investors alerted the Shenzhen authorities to the value of Futian, so they broke the leases and took back the land to develop it into a bonded area. The downpayments were returned to the investors without interest after long delays. But the development of Futian was far below expectations, partly because of the excessive fees levied by the Board. Another problem was that the Board formed companies of its own to compete with the private investors. For instance, after private investors realized that freight consolidation is profitable in Futian, the Board formed its own consolidation company.

The problem is that Shenzhen has monopoly power because it is the only locality in the Pearl River Delta immediately accessible by land from Hong Kong, and some officials are understandably more interested in exploiting this monopoly power to their own advantage than pursuing long-term economic development.

In this light, the proposed Zhuhai-Hong Kong bridge is important not only in terms of transportation, but also in terms of the political economy. The bridge will break the monopoly power of Shenzhen.

Shenzhen as a Bonded Area

Shenzhen has plans to become a huge bonded area or free trade zone by 'opening the first line and tightening the second line'. However, the plan is fraught with technical difficulties that will have highly undesirable consequences.

It will be very difficult to prevent the smuggling of people and goods between Shenzhen and the rest of China. Even if these technical difficulties

2. *Economic Integration of Hong Kong and the Zhujiang Delta*

can be overcome, the tightening up of the second line implies that the people and goods going from Hong Kong to China via Shenzhen will have to go through customs twice. Moreover, the rest of China stands to lose because its links with Shenzhen will be weakened.

There is now rampant smuggling between the mainland and Hong Kong despite strenuous efforts to crack down on the illegal traffic on both sides of the border. Once Shenzhen becomes a free trade zone, luxury consumer goods will enter Shenzhen freely and Hong Kong's smuggling rings will move operations to Shenzhen. Policing will be extremely difficult because the length of the second line is thrice the length of the Hong Kong-Shenzhen border, and the length of the coast of Shenzhen is roughly the same as that of the second line (*Hong Kong Economic Daily,* 24 and 25 April 1992). Moreover, the second line has been poorly designed; it goes through some built-up areas and has no roads for patrolling in many places. Smugglers will be able to easily cut the fences and walk through.

Presently, smugglers have to deal with the police forces of two different systems, namely of Hong Kong and China. Once the smuggling rings move to the bonded area of Shenzhen, they will only have to deal with or bribe the police of one system. In Shenzhen's August 1992 sale of application forms for new share subscriptions, nearly a million people rushed into Shenzhen, resulting in demonstrations and riots. This episode shows that the second line is powerless to stem would-be immigrants.

After Shenzhen becomes a free trade zone, goods going from Hong Kong to the area north of Shenzhen will have to go through customs twice: once when the goods go from Hong Kong into Shenzhen, and a second time when they go through the second line. The economic losses of going through customs twice are considerable. According to a survey of the Hong Kong General Chamber of Commerce in 1992, Hong Kong trucks going into Shenzhen on average have to wait an hour, resulting in losses of HK$30 billion per year (*Hong Kong Economic Times*, 22 May 1992).

Presently, people leaving Shenzhen through the second line do not have to go through customs, but trucks leaving Shenzhen to go inland still have to go through customs. This is because Shenzhen can import selected amounts of consumer goods for use in Shenzhen at half the official tariff rates. As the quantity of goods entitled to partial tariff exemptions is strictly controlled, the problem of smuggling is not serious and trucks do not have to line up to pass the second line.

Once Shenzhen becomes a free trade zone, both people and trucks leaving Shenzhen to go inland will have to be inspected, and congestion at the second line will be a major problem. Some commentators have opined that the containers of Hong Kong's trucks can be inspected and sealed at the Hong Kong-Shenzhen border, and then there would be no need for another inspection at the second line. However, there are plenty of areas on a container truck, such as the driver's compartment, that cannot be sealed. It is difficult to avoid a second inspection. The only way to avoid a second inspection is to build a sealed highway through Shenzhen, but the feasibility and cost of such a project have yet to be studied.

The undesirable consequences of creating the huge bonded area of Shenzhen are nonexistent in Futian or in other bonded areas. Hong Kong trucks going north into China have to pass through Shenzhen, but they do not have to go through Futian or other bonded areas, and the problem of going through customs twice will not arise. Moreover, Futian or other bonded areas do not have permanent residents. Therefore, strict controls on visitors to stem illegal immigration will not have to be imposed. Similarly, smuggling rings will not be able to use them as bases of operation.

Lastly, turning Shenzhen into a bonded zone risks the animosity of the inland areas. Beijing's official policy is to let selected areas 'get rich first' and then to encourage laggards to catch up. However, if the favoured area encloses itself with a 'second line' after getting rich and then imposes strict controls on the immigration of poor cousins, the policy will weaken the diffusion of the developmental process from the developed areas to the less developed areas. The policy could easily lead to economic polarization.

Border industrial parks or development zones

Due to the shortage of land in Hong Kong, there have been proposals to establish an industrial park in Shenzhen just adjacent to Hong Kong. Besides cheap land, the park would be able to tap the labour and engineering skills of China as well as the efficient supporting services of Hong Kong.

However, an industrial park in Shenzhen will have to be managed by the Shenzhen authorities and the China Customs Administration, which are not known for their efficiency. In fact, under efficient management, Futian could function as such an industrial park. But since Futian did not prosper, there is little hope that the proposed park would prosper.

The alternative would be to establish a park at the border areas of Hong Kong. The park would then be under Hong Kong management. There was such a proposal from textile and clothing manufacturers some years ago. The purpose was to import cheap labour from China to utilize Hong Kong's textile and clothing quotas. The proposal was turned down by the Hong Kong government as it would have provided excuses to importing countries to cut Hong Kong's quotas. In any case, since the MFA will be phased out by 2005, there is little reason to rescue a sunset industry.

Apart from the textile and clothing industries, there are some reasons to look into the establishment of an industrial park in the border areas of Hong Kong for more advanced industries. Besides cheap labour, China also has a vast army of engineers and considerable strength in research and development. Hong Kong has ample land in the border areas as it has maintained a vast band of undeveloped land along the border to strengthen border control. With miraculous economic development in Shenzhen, there have been few desperate refugees risking their lives to scale fences to enter Hong Kong. Illegal immigrants have used safer channels such as high-speed motor boats. It is time for the Hong Kong government to reconsider land use in the border areas.

The "One Country, Two Systems" Research Institute has proposed turning Hong Kong's border area into a huge development zone that would allow freer access of mainlanders. One problem with the proposal is that Hong Kong would have to make sure that its "second line" south of the border zone could be effective in preventing illegal immigration. The other problem is one of international image. A border industrial park, or a grand border development zone allowing freer access of mainlanders, may affect international perceptions of Hong Kong as an autonomous and unique region distinct from China. It may be difficult to persuade foreign governments that advanced technology, especially technology with military applications, that is imported by Hong Kong will not leak to the mainland. The functioning of "one country, two systems" presumes strict and effective border controls.

Barriers to immigration

Hong Kong residents have a large number of family members on the mainland, especially in the Zhujiang Delta. Migration to Hong Kong is controlled by a quota of 150 immigrants per day or 54,750 immigrants per year. This quota is a very high barrier to economic integration. Spouses of Hong Kong residents often have to wait for ten or more years before they can migrate to Hong Kong.

Cross-border marriages have become increasingly popular since China's open policy. A 1999 Hong Kong government survey estimated that Hong Kong residents have 794,000 children in China (including 520,000 born out of wedlock), of whom around 267,600 (including 170,000 born out of wedlock) have the right of abode in Hong Kong according to the Basic Law (i.e., at least one of their parents has to be a permanent resident at the time of the child's birth) (Census and Statistics Department, 1999).

According to calculations, within five years Hong Kong will absorb these 267,000 children who have the right of abode. However, their population will keep rising over time, partly because more children will be born to existing couples, and partly because there will be more cross-border marriages. Moreover, part of the quota will have to be used for the migration of spouses. As a result, the absorption of children who have the right of abode may take forever.

Presently, due to fears of population pressures, public opinion is in favour of strict controls on migration, not only for family reunions, but also for children who have a constitutional right of abode in Hong Kong. This mentality is best termed as "Superfortress Hong Kong". However, the mentality and logic of "Superfortress Hong Kong" are inconsistent with Hong Kong's long-term integration with the Zhujiang Delta. As a result of the high costs of living in Hong Kong, more and more Hong Kong residents have moved to Shenzhen and they commute to Hong Kong to work. At the end of April 2000, the Hong Kong government estimated that there were 40,000 Hong Kong residents who had moved to Shenzhen (*Ming Pao*, 27 April, 2000). A survey of the Hong Kong-China Relation Strategic Development Research Fund revealed that up to one million Hong Kong residents may consider relocating to Shenzhen in the next decade or two.

In the long run, it is inevitable that a large number of Hong Kong residents will relocate to Shenzhen and commute to Hong Kong to work. However, such a scenario is inconsistent with "Superfortress Hong Kong". On the one hand, planners envisage one million Hong Kong residents living in Shenzhen and commuting to Hong Kong to work every day. On the other hand, if these commuters marry mainlanders, their spouses and children will have to wait forever to obtain permission to enter Hong Kong. Government officials simply cannot plan when, where and with whom individuals will marry.

It is clear that "Superfortress Hong Kong" does not make sense in the long run, and Hong Kong's present stringent controls on migration have to be relaxed, though some controls must remain in order for the "one country, two systems" to remain viable. Paradoxically, the ultimate solution to population pressures in Hong Kong lies in deeper integration with the Zhujiang Delta. Hong Kong can encourage people to relocate to Shenzhen by expanding the capacity of border crossings to facilitate commuting. Hong Kong can also encourage retired people to relocate to the Zhujiang Delta by building housing estates designed for retirees there.

Opportunities for economic co-ordination

Instead of old-style integration through discriminatory preferences, it is more fruitful to concentrate on new-style integration through policy co-ordination. Such co-ordination does not contravene the obligations of the mainland and Hong Kong to the WTO. It is also less likely to affect international perceptions of Hong Kong as an autonomous entity. Moreover, such co-ordination is consistent with the reputation of Hong Kong as a free economy open to the whole world.

Hong Kong's reversion to Chinese sovereignty in 1997, the AFC, and China's prospective entry into the WTO have promoted policy co-ordination between China and Hong Kong. Since 1997, policy co-ordination has moved forward in seven areas, namely, border area issues, regional infrastructure, tourism, technology policy, financial markets, stabilization of the exchange rate, and China's entry into the WTO.

Border area issues

Border area issues involve border checkpoints, environmental protection, and co-ordination of cross-border infrastructure. These issues involve co-ordination with Guangdong, though the cross-border infrastructure involves the State Council (the Ministry of Transport and the Planning Commission) as well.

A Hong Kong/Guangdong Co-operation Joint Conference was established in March 1998, headed by Hong Kong's Chief Secretary and Guangdong's Vice Governor. In September 1998, the Joint Conference reached agreements on environmental protection, on minor extensions of the working hours of border checkpoints, and on measures to boost cross-border tourism. The Joint Conference was supposed to meet twice a year. However, it has not met since September 1998 due to a lack of items for discussion. This is a sign that the Hong Kong government has not been very active in promoting co-ordination with Guangdong.

Environmental protection

To protect the environment, the two sides agreed to subject all large-scale public works in the border areas to environmental evaluation and mutual consultation. The two sides undertook a joint study of air quality in the Pearl River Delta in 1999 as pollutants in the Delta have led to a substantial deterioration of air quality in Hong Kong.

Border checkpoints

Congestion at border checkpoints has long been a source of complaints. Passenger crossings often take an hour or more, and as a result many commuters are affected. There were proposals to allow 24-hour operation of the major passenger checkpoints after Hong Kong's reversion, but the agreement reached in late September 1998 after long negotiations was disappointing as the opening hours of the four land crossings were extended by only one to two hours. Even after the extensions, there were no checkpoints open from 11.30 p.m. to 6.30 a.m.

The Hong Kong Government has reservations about 24-hour operation of passenger checkpoints. This is because retail sales in Hong Kong have plummeted since the AFC partly because Hong Kong shoppers have swarmed to Shenzhen, where prices of goods and services are much lower. Greater ease of passenger crossings will have an adverse impact on retail sales and employment in Hong Kong. It will also depress real estate prices in north New Territories.

Checkpoint congestion is a barrier to trade in services. It is ironic that Hong Kong, which is known internationally as a bastion of free trade, implicitly relies on a trade barrier to protect its employment and other vested interests. However, the public outcry against checkpoint congestion was so strong that the government appeared to be preparing for 24-hour operation of the Lok Ma Chau checkpoint in May 2000.

Border tourism

Hong Kong and Guangdong reached an agreement on five measures to boost cross-border tourism at the Joint Conference of September 1998. Besides facilitating mainlanders' visits to Hong Kong, the measures also increased the attraction of Hong Kong to overseas visitors as there were many more places they could visit in Guangdong without a visa. Moreover, Hong Kong tourist agencies can share in part of the tourist business involving overseas tourists visiting Guangdong as well as Guangdong tourists visiting Hong Kong or elsewhere. Of course, Guangdong would also benefit from an increase in overseas tourists.

Regional infrastructure

It has been pointed out that there are too many seaports and airports in the Zhujiang Delta. Within just a few years after Deng's 1992 southern tour, Beijing approved the construction of three deep water ports adjacent to Hong Kong, namely, Yantian in Shenzhen, the Huizhou port in Daya Bay, and Gaolan port in Zhuhai. This is in addition to the existing Guangzhou port and Shekou port in Shenzhen (Sung, Liu, Wong and Lau, 1995: 193–198). Though the Hong Kong port is congested and there is a need for another port in the Zhujiang Delta, there will be inadequate freight to supply so many deep-water, ocean-going ports.

A successful international port must have sufficient freight to attract shipping companies. A large and busy port tends to be efficient because of the economies of agglomeration: An increase in freight will result in a more frequent shipping schedule which in turn will attract more freight to the port as the freight can be shipped out speedily. The process snowballs until the port is congested. Then there is a need to develop a new port.

Hong Kong has had the busiest container port in the world for some time, handling 16.2 million containers in 1999. A minimum freight volume of 1.5 million containers a year is needed for the efficient operation of a container port (Sung, Liu, Wong and Lau, 1995: 196). All the Chinese ports near Hong Kong have quite a long way to go to reach the threshold of 1.5 million containers a year, with the exception of Yantian, which handled over a million containers in 1998. The Huizhou and Gaolan ports have yet to take off, though some berths have already been completed. By spreading its efforts too thin in too many ports, China is making it difficult for any port to compete with Hong Kong.

Like port construction, there is also duplication in airport facilities. Within a region of 200 square kilometres in the Pearl River Delta, there will be five international airports (Hong Kong, Macau, Shenzhen, and two airports in Guangzhou) and four local airports (Sung, Liu, Wong and Lau, 1995: 198). Each local government looks after its own interests and wants a seaport or an airport as a showpiece. The result is excess capacity and duplication of facilities.

Co-ordination and competition

Though improvements in planning and co-ordination can avoid some duplication of facilities, a certain amount of excess capacity and duplication is essential for competition. Planners may be able to rule out the more obvious cases of failure, but they may not have enough information to pinpoint the successful cases. For instance, the choice of Shekou as an ocean-going port is problematic because of its shallow waterway which needs dredging. However, planners are probably not able to decide between Yantian and Gaolan, as both have a good potential. A certain amount of trial and error is necessary and no amount of feasibility studies can replace the market tests. The problems with ports and airports in the Pearl River Delta is not excess capacity *per se*, but the ability of local governments to subsidize them from the public budget. Public subsidies imply that there is no real market test as loss-making ports or airports can continue to operate.

It has been pointed out that the participation of foreign capital in infrastructure development will imply more efficient use of resources as foreign investors will help government officials evaluate the cost-effectiveness of infrastructure projects. The over-development of local airports in the Pearl River Delta has been attributed to the lack of participation of foreign capital (Sung, Liu, Wong and Lau, 1995: 202).

Co-ordination of infrastructural development after 1997

The Sino-British dispute over the Patten proposals for constitutional reform in Hong Kong hindered co-ordination of infrastructure development between Hong Kong and China. For instance, both the construction of Hong Kong's new airport and of container terminal number 9 were held up for a while due to the Sino-British dispute. The reversion of Hong Kong to China has led to better co-ordination of infrastructural developments in the Pearl River Delta. A Hong Kong-Mainland Cross-Boundary Major Infrastructure

Co-ordinating Committee was established in October 1997, taking over the agenda of its predecessor established in December 1994.

However, it should not be assumed that Hong Kong's reversion will solve the problem of co-ordination. For instance, neither Beijing nor the provincial government of Guangdong has been able to rationalize the many airports and seaports put forward by the local governments of the Pearl River Delta. In Shenzhen alone, there are two ocean-going ports, Yantian and Shekou, which is one too many. The Shekou Industrial Export Zone was formed one year before the Shenzhen SEZ and was managed by the China Merchants Company based in Hong Kong. Though Shekou was incorporated into Shenzhen, it continued for a while to be administered by the China Merchants Company. Shenzhen was unable to stop the construction of the Shekou port.

Bureaucratic coordination may not be powerful enough to override local interests. Even if bureaucratic coordination is powerful enough, it is unlikely to be highly rational and efficient. Market coordination will provide better results as long as the local governments refrain from giving undue subsidies to infrastructure. The participation of private capital and foreign investors in infrastructure development should be encouraged as much as possible. Recently, China has allowed increased participation of foreign capital in infrastructure projects, including seaports. Hong Kong International Limited (HIT) has invested in the ports of Yantian, Shanghai, Gaolan, and other feeder ports in the Zhujiang Delta (Cheng and Wong, 1997: 61). Swire Pacific and Peninsular & Oriental Stream Navigation Co (P&O) have acquired 50 percent of the Shekou Container Terminal.

The rapid growth of Yantian is largely due to efforts by HIT. The design capacity of Yantian reached 1.7 million containers in 1999. However, Yantian needs time to grow and Hong Kong will remain the major hub for years to come, with Yantian serving as a subsidiary port (Cheng and Wong, 1997: 62–68). Despite higher costs, the Hong Kong port is very efficient due to its very frequent shipping schedules. For instance, in 1999 Hong Kong had 450 weekly container services to 170 ports in 60 countries whereas Yantian had only 13 weekly services.

Though container throughput stagnated in Hong Kong in 1998 due to the AFC and Hong Kong lost its position as the world's busiest container port in 1998 to Singapore, Hong Kong controlled its costs and regained the number one position in 1999. While the share of China's cargo going through

Hong Kong will diminish, the absolute volume of cargo handled by Hong Kong will continue to rise.

Tourism

Hong Kong's tourist industry was severely hit by the Asian financial crisis because the Hong Kong dollar did not depreciate. At the end of May 1998, the Government announced a package of seven measures to stimulate the sagging economy. One of the measures involved a 30 percent increase in the quota allocated to mainlanders' joining the Group Tour Scheme to visit Hong Kong. (Such tours were started in the early 1980s and the quota was instituted to avoid illegal immigration.) Another measure was to simplify visa formalities for Taiwanese tourists who tour Hong Kong on their way to the mainland. Such measures brought an upturn. The agreement with Guangdong in September 1998 mentioned above to promote cross-border tourism also helped.

Hong Kong Disneyland, scheduled to open in 2005, is projected to attract 3.4 million overseas tourists. By 2020, the Park will reach full capacity, attracting 7.5 million overseas tourists. The bulk of the overseas tourists will be from the mainland and this will necessitate further coordination between Hong Kong and China.

Technology policy

Despite Hong Kong's lagging technical capacity in comparison with Singapore, Taiwan, and South Korea, Hong Kong is uniquely positioned to utilize the research and engineering capability of the mainland. Moreover, in comparison with its rivals, Hong Kong has the advantage of close links with the mainland's vast internal market. The Government announced new measures to promote technology in October 1998, including the establishment of an applied science and technology research institute. The importation of skilled professionals from the mainland without quotas started in early 2000.

Upgrading Hong Kong's technology is perceived to be in the interest of Guangdong and the mainland as the labour-intensive processing industries in Guangdong need upgrading. In July 1999, China's State Council designated Shenzhen as a pilot hi-tech development area to jointly promote hi-tech with Hong Kong. Beijing has already approved the setting up of around 50 hi-tech institutions in Shenzhen, where they can cooperate with Hong Kong businesses.

Financial markets

In 1992 China approved the public listing of selective state enterprises on the Hong Kong stock market. Their shares are called H-shares. Listing in Hong Kong speeds up the mainland's enterprise reforms as listed firms have to follow international accounting standards. Just before the reversion in May 1997, H-shares comprised 24 companies, with a market capitalization of US$7 billion or one percent of Hong Kong's market. The number of H-shares rose to 39 by October 1997 when the AFC hit Hong Kong. By April 2000, there were 45 H-shares. Growth after the AFC was slow.

Besides H-shares, Red Chips or stocks of listed Hong Kong companies controlled by mainlanders are also important. In May 1997, there were around 50 Red Chips, with a market capitalization of US$49 billion (10 percent of Hong Kong's market). A China-Affiliated Corporations Index or Red Chips Index was introduced on 16 June 1997 as such shares were very popular.

The AFC ended the Red Chips bubble. Abuses and mismanagement of China's companies, as epitomized by the 1998 closure of the GITIC (Guangdong International Trade and Investment Corporation), the second largest corporation owned by Guangdong, continued to dampen investor interest in H-shares and Red Chips. However, in early 2000 China's expected entry into the WTO revived some interest in Red Chips.

Despite China's problems with SOE (state owned enterprise) reform, China's rapid development implies a huge demand for foreign funds (Liu, 1998: 23). As more Chinese enterprises seek listing in Hong Kong, Hong Kong may become the world's largest stock market within several decades.

While Hong Kong's comparative advantage in funding China's development may help Hong Kong develop into a truly first rate international financial centre, managing the Hong Kong currency will be increasingly difficult as it is used for the transaction of the shares of Chinese enterprises listed in Hong Kong. The volume of such transactions will be exceedingly large and the Hong Kong currency will be vulnerable to speculation. In the long run, it is better to trade such stocks in US dollars to insulate the local economy from the flow of funds involved in offshore financial activities (Liu, 1998: 23–24).

The Hong Kong Monetary Authority has secured the help of HSBC, a private bank, to assume the risk of clearing US dollars in Hong Kong. A RTGS (Real Time Gross Settlement) clearing system in US dollars will be established in the third quarter of 2000. This will promote US dollar-based financial transactions in Hong Kong and also enhance Hong Kong's status as an international financial centre. Presently, the bulk of syndicated loans and bonds in the Hong Kong market are denominated in foreign currencies (mostly US dollars). However, stocks are quoted and traded in Hong Kong dollars, though more and more are likely to be quoted and traded in both US and HK dollars in the future.

The establishment of US dollar clearing facilities in Hong Kong amounts to a partial "dollarization" in which large financial transactions will be conducted in US dollars while daily transactions will still be conducted in Hong Kong dollars (that is linked to the US dollar at a very stable exchange rate). Such a partial dollarization would not be possible without the implicit blessing of China.

A Second Board was established in Hong Kong in November 1999 to facilitate the raising of capital by new companies. Many new technology companies raise capital through the Board, including technology companies in China.

Co-ordination to stabilize the exchange rate

Unlike other East Asian economies, the mainland and Hong Kong stood firm and maintained their exchange rates against the US dollar during the AFC. Both governments have reiterated their determination to keep their exchange rates stable. Co-ordination in exchange rate policies between the mainland and Hong Kong have contributed to financial stability in East Asia (Sung, 1998: 165–167). President Clinton acknowledged the importance of the economic roles of the mainland and Hong Kong in the Asian financial crisis during his visit to China and Hong Kong in late June and early July 1998.

Unlike Hong Kong, the mainland's long-run interest is best served by a flexible rather than a fixed exchange rate. Hong Kong benefits from a fixed exchange rate much more than the mainland because Hong Kong is a small economy where international transactions are relatively much more important. A fixed exchange rate system facilitates international transactions. While many small economies have pegged their currencies to those of their major economic partners, most large economies allow their exchange rates to float. For large economies, the burden of a fixed exchange rate system usually outweighs the gains. However, during the AFC, exchange rate stability served the interest of China as well as that of its neighbours. East Asian financial markets were very unstable in 1998, and a devaluation of the renminbi would have triggered another round of devaluations and financial turmoil in East Asia. China had relatively little to gain from devaluation as long as financial markets in East Asia were unstable.

As financial stability has returned to Asia, China can allow more flexibility in its exchange rate movements. After China's entry into the WTO, China's huge surplus in trade may shrink due to import liberalization. There are signs that China may allow more flexibility in its exchange rate to cope with WTO entry.

Co-ordination on China's WTO entry

Hong Kong has long supported China's entry into the WTO. Hong Kong government officials and politicians also lobbied the US government and the US Congress to grant NTR (Normal Trading Relations) status to China in early 2000.

In his Policy Address of October 1999, Tung Chee-hwa announced the setting up of an inter-departmental group chaired by the Financial Secretary to liaise with mainland authorities on matters related to China's WTO entry. The purpose is to become acquainted with arrangements for the opening up of the mainland market, and to enable Hong Kong businesses to capitalize on the emerging opportunities. Additionally, China's MOFTEC (Ministry of Foreign Trade and Economic Co-operation) and Hong Kong's Trade and Industry Bureau set up a joint committee in November 1999 to strengthen communication on economic and trade issues.

A "new-style" FTA?

The conclusion of the Sino-US agreement on China's WTO entry revived talks of a mainland-Hong Kong FTA and the Hong Kong Chamber of Commerce proposed in early 2000 a "new-style" FTA concentrating on services trade. However, in comparison with those of the mainland, Hong

Kong's service sectors are very open. It would be very difficult for the mainland to match Hong Kong's openness in services in a FTA agreement.

China's priority is to enter the WTO and open its service sectors on an unprecedented scale. This will impose a severe adjustment burden on China's inefficient service sectors. Until China overcomes its adjustment problems, it is not practical to ask China to open up its service sectors further to Hong Kong.

In the long run, a "new-style" mainland-Hong Kong FTA is not without merits. After China has entered the WTO and has overcome its adjustment problems, China may wish to experiment with further liberalization of its service sectors on a selective basis. In that case, a "new-style" mainland-Hong Kong FTA can be explored.

It should be noted that a FTA has obvious political implications in addition to its economic impacts. A mainland-Hong Kong FTA will of course symbolize the union of Hong Kong to the motherland. It is conceivable that political circumstances may favour a Taiwan-Hong Kong FTA, or even a trilateral FTA composed of the mainland, Taiwan, and Hong Kong.

As Japan is considering a Japan-Singapore FTA, Japan may also be interested in a Japan-Hong Kong FTA which will strengthen Japan's links with the Chinese Economic Area. Of course, such a FTA will require the blessing of Beijing. As far as Hong Kong is concerned, it is advantageous to be involved simultaneously in FTA agreements with the mainland and other friendly East Asian countries. Hong Kong should strengthen its international links as well as its links with the mainland.

Conclusion

Despite the huge investments by Hong Kong in the Pearl River Delta, economic integration between the two areas has been highly uneven due to the many differences in their political, legal, and economic systems. The integration of export-oriented industries (outward processing) has developed extremely rapidly because such products are exported to the world market, and they are not hampered by China's foreign exchange controls. Foreign investments in industries selling to China's domestic market face more problems due to China's foreign exchange controls.

Similarly, the integration of service industries is slow because most services are sold on the domestic market and cannot be exported. Moreover, high-end services, such as financial services, telecommunications and the information industry, are highly regulated, and China's regulatory regime is cumbersome and non-transparent.

Despite the many barriers to integration, Hong Kong has been able to achieve a foothold in many service sectors as a result of its vast business network that grew from their pioneering investments in outward processing.

China's entry to the WTO will imply a more transparent regulatory regime which will speed up the integration of services. This will benefit Hong Kong as it has a substantially stronger comparative advantage in services. Though Hong Kong's service providers will face competition from multinational service providers, Hong Kong already has a headstart due to its established lead in Chinese trade and investment. Moreover, Hong Kong has the advantage of cultural and geographic proximity.

The reversion of Hong Kong has facilitated policy co-ordination between Hong Kong and the mainland. Despite the many joint committees established, the present co-ordination is quite *ad hoc* and reactive, lacking in overall vision and strategy. For instance, 24-hour operation of passenger checkpoints is long overdue.

As mentioned above, the policy of "Superfortress Hong Kong" is not consistent with long-run integration with the Zhujiang Delta. Hong Kong has to make hard choices and gradually relax the very stringent controls over migration of mainlanders. There will be burdens on Hong Kong's social welfare and social services and also pressures on unemployment. However, Hong Kong can relieve such pressures through reforms of social welfare and social services. For instance, Hong Kong can impose length of residency requirements on eligibility for social welfare and social services. It is preferable to reduce the economic incentives of migrating to Hong Kong and let eligible individuals choose where to locate their families because the present stringent controls on migration impose huge social and economic costs and also disrupt family life.

The Policy Address of our Chief Executive, Mr Tung Chee-hwa, in October 1999 included sections on "Strengthening Ties with the Mainland", and "Joint Development of the Pearl River Delta Region", but there was no articulation of an overall vision of future integration. An articulation of such a vision is important because it would set the basis of a community-wide debate which would raise an awareness of the far-reaching implications and potential of deep integration with the mainland.

Lastly, officials in Hong Kong should closely watch the many "new-style" FTAs that have been proposed in APEC as they have obvious implications for the Chinese Economic Area. Though these "new-style" FTAs are presently no more than speculative visions, it is conceivable that Hong Kong may be involved in such FTAs with the mainland, Taiwan, Macau, or other APEC members in the future.

References

Census and Statistics Department. *Hong Kong Residents with Spouses/Children in the Mainland of China*. Special Topics Report No. 22. Hong Kong: Government Printer, 1999.

Cheng, Leonard K., and Wong, Yue-Chim Richard. *Port Facilities and Container-Handling Services*. Hong Kong: City University of Hong Kong Press, 1997.

Liu, Pak-wai. *The Asian Financial Crisis and After: Problems and Challenges for the Hong Kong Economy*. Occasional Paper No. 89, Hong Kong Institute of Asia-Pacific Studies, The Chinese University of Hong Kong, December 1998.

Sung, Yun-Wing. *Hong Kong and South China: The Economic Synergy*. Hong Kong: City University of Hong Kong Press, 1998.

Sung, Yun-Wing, Liu, Pak-Wai, Wong, Yue-Chim Richard, and Lau, Pui-King. *The Fifth Dragon: The Emergence of the Pearl River Delta*. Singapore: Addison Wesley, 1995.

PART II

Urbanization and Regional Development

Fertility of Migrants and Non-migrants in Dongguan and Meizhou: A Study of the Impact of Regional Development and Inter-regional Migration on Fertility Behaviour in China

Si-ming Li and Yat-ming Siu
Department of Geography, Hong Kong Baptist University

Introduction

It is a truism that China is enduring intense population pressures. With a population exceeding 1.2 billion, China is the world's most populous country. Climatic and topological factors set strict limits on the amount of land that can be cultivated. No more than 10 percent of the country's 9.6 million square kilometres of land is cultivable (Ketizu, 1992). That is, in China each hectare of farmland has to support, on average, some 13 people. Recently, because of rapid and often haphazard urban development, China has been losing substantial amounts of its most productive farmlands (Yeh and Li, 1999). This is especially the case in the rich alluvial plains of the Yangzi and Pearl River deltas and in the Bohai region. In light of all this, it is indeed a remarkable accomplishment that to date China has been able to feed herself, primarily through consistent and steady increases in agricultural productivity over the past forty years (Brown, 1995). But this has only been achieved by massive increases in the use of chemical fertilizers and

This research is supported in part by Hong Kong Research Grant Council research grant number HKBU 2080/99H. The authors would like to thank Miss Doris Fung for assistance in data processing and the Population Research Institute of the Guangdong Provincial Academy of Social Sciences for help with the field survey.

insecticides, which obviously have serious environmental consequences. There are limits to how much further agricultural productivity can be raised in the future, given the level of technology available to Chinese farmers and the willingness or unwillingness of the state to allocate additional resources to the agricultural sector. Considering the size of China's population in relation to the world's total, quite naturally the question of who can feed China if she cannot do the job herself has become a hot topic for debate both within and beyond academic circles (Brown, 1995).

Under such circumstances, it is not surprising that population control, in particular birth planning, has been a cornerstone of China's economic and social policy for more than thirty years. Indeed birth planning is enshrined in the Constitution (Siu, 1995; 2000). With strict enforcement of the one-child policy, which has been in place since 1980, and other related measures such as deferred marriages and forced abortions, the crude birth rate decreased from 21.04 per thousand in 1985 to 16.03 per thousand in 1998. Correspondingly, the natural population growth rate decreased from 14.26 per thousand to 9.53 per thousand over the same period (SSB, 1999, p. 112). It has been estimated that China would have had 200 million extra births in the past twenty years had there not been strict implementation of birth control measures (Wei, 1998). With economic growth and urbanization gathering pace, it may be expected that the crude birth rate, and hence the population growth rate, will show further declines, judging from the experience of other countries (Siu, 2000; Teitelbaum, 1975).

Yet it has been alleged that the increasing incidence of population migration brought about by sharpening regional differences and urban and rural disparities, and by the reforms in the urban economic sectors, including the gradual relaxation of the *hukou*, or household registration system (Siu and Li, 1993), have rendered birth planning difficult. More specifically, the authorities are faced with immense difficulties to keep track of those women who have moved away from their home villages or towns to seek fortunes largely in the major cities in the southeastern coastal region. Given China's household registration system, two types of migration may be identified, namely, *de jure* or permanent migration and *de facto* or temporary migration (Chan, Liu, and Yang, 1999; Li and Siu, 1997; Li, 1997). The former refers to migration with a corresponding change in household registration, and the latter refers to migration without a change in registration. Whereas *de jure* migration is state-sanctioned and tightly controlled, *de facto* migration is legally dubious and largely spontaneous. Statistical evidence indicates that the latter has experienced phenomenal increases in recent years (Siu and Li, 1993). The 1995 National 1% Population Sample Survey puts the size of *de facto* migration at 49.7 million (SSB, 1997a, pp. 538–39). Not surprisingly, Guangdong, Jiangsu, Beijing and Shanghai are among the most popular destinations of the temporary migrants.

Because of the transient nature of the temporary migrants, it is difficult for the authorities to implement family planning and enforce birth control measures on them. For example, it is unclear whether the responsibility of birth planning rests upon the local authorities at the place of origin or on those at the place of destination. Theoretically, it should be the former, as the place of household registration remains unchanged. However, these migrants currently reside in places hundreds or even thousands of kilometres away from their home villages. It is simply impossible for the home village authorities to extend their jurisdiction to the migrants' current places of residence. As for the local authorities at the place of destination, their overriding concerns are whether the migrants' presence can contribute to the local economy and whether there are criminals within the migrant population. Birth planning for the migrants is beyond their jurisdiction. In any case, the authorities at the host destination do not have to take care of the welfare of the migrants and their offspring. There are no incentives for them to oversee their birth planning. Thus, it is widely believed that the increasing incidence of migration, especially *de facto* or temporary migration, has a negative effect on birth control enforcement. Temporary female migrants of childbearing age are often called "excess births guerrillas" (*chaosheng youjidui*) (Gui, 1992; Zhou, 1996).

Based on data obtained from two household surveys conducted in 1992, one in Dongguan and the other in Meizhou, both in Guangdong province, this study attempts to answer two interrelated questions concerning fertility behaviour in reform China. First, there is the question of whether or not, and to what extent, fertility behaviour in China under the market-oriented reforms is related to the level of economic development. Second, there is the question of whether migration, especially *de facto* migration, which is closely related to differential economic growth between regions, affects fertility propensity. This latter question assumes special importance not only in light of the massive scale of inter-regional migration in China but also in light of China's unique household registration system. Dongguan, located in the Pearl River Delta, and Meizhou, located on the eastern fringe of the province, were chosen primarily because of their contrasting economic performance since the launching of the reforms. A comparison of fertility behaviour in these two places provides hints to the answer to the first question. Dongguan is also known to have attracted huge numbers of *de facto* or temporary migrants (Fan, 1996; Li and Siu, 1998). The survey contains rather detailed information about the respondents' household and migratory status. This will shed light on the second question. In a sense the present paper is an extension of Siu's (2000) previous work, which focuses primarily on the first question. The main contribution of the present paper is the inclusion of migration variables in the study of fertility behaviour, which seems to have been neglected in the literature.

The data

The two sample surveys which provide the data for the present study were carried out by the authors in conjunction with the Population Research Institute of the Guangdong Provincial Academy of Social Sciences in 1992.

Dongguan, primarily because of its proximity to Hong Kong, is a major beneficiary of China's open-door policy. In 1983 when first accorded municipal status, Dongguan's landscape was similar to that of other rural counties in the Pearl River Delta, with only scattered urban settlements including the county seat and a number of rural towns dotting the landscape. Today, however, this emerging and sprawling metropolis is filled with tens of thousands of factories mainly engaged in export trade financed by Hong Kong and, to a lesser extent, Taiwan capital. In 1998, Dongguan's total exports amounted to US$13.06 billion, or some 17.3 percent of the provincial total (SBG, 1999, p. 608). These factories, together with thousands of housing construction projects primarily targeting the Hong Kong market, and the service industries generated by the multiplier effect, employ over a million workers, the bulk of whom are *de facto* or temporary migrants from all over the country (Li and Siu, 1997; 1998). The 1995 National 1% Population Sample Survey shows that Dongguan had 467,000 *de facto* migrants in that year, as compared with a population of 1.026 million with proper residential status (SSB, 1997b, pp. 250–251). But even this figure is likely to be a gross underestimation of the extent of temporary migration, as most foreign-owned factories customarily do not reveal the actual number of migrant workers to the authorities so as to minimize the payment of fees, such as the city accommodation charge (*chengshi zengrong fei*) and the temporary residence registration fee (*zanzhu hukou dengji fei*).

Meizhou is located in the hilly region in the northeastern corner of Guangdong province. Although the city has been able to report respectable growth since the launching of the reforms, its ability to attract foreign capital lags behind that of Dongguan by a huge margin. The per capita GDP of Meizhou in 1998 was only RMB 3,231,[1] which was no more than a fraction of Dongguan's RMB 24,031 (SBG, 1999, p. 109). But Meizhou is the centre of the country's tens of millions of Kejia (Hakka) people, a language group of people in China known for their emphasis on education. According to the 1995 1% National Population Sample Survey, 24.06 percent of Meizhou's population aged 6 or above have received senior secondary education or higher. In comparison, the figure for the much more prosperous Dongguan is only 9.56 percent (SSB, 1997b, pp. 90–99). Reflecting its relative importance in the cultural dimension, Meizhou also has quite a large number of migrants, but these migrants are primarily of the *de jure* type and migration is often related to marriage or job assignment by the state (Li and Siu, 1997).

At time of the survey (1992) large parts of both Dongguan and Meizhou still maintained a largely rural character, despite their official municipal status. In the survey, efforts were made to differentiate the former county seats from the market towns, and the urban settlements from the rural areas. In both cities, some 800 observations were taken based on the household registration records. In each city, the observations were composed of the following: approximately 250 were from the former county seats; the rest, approximately 550 each, were from two townships. In each township, about 75 were from the town seat, and 250 were from the rural area. To take into account the large number of temporary or *de facto* migrants in Dongguan who, by definition, would not be recorded in the book of household registration, a supplementary sample of 405 households in this city was taken, using the registry of temporary residents as the sampling base. The survey instrument, i.e., the questionnaire, contains more than 150 questions covering each member of the responding household's residential and migration history, household registration status, current and previous jobs, household and personal incomes, and, for married women, a detailed account of their fertility behaviour. The section on fertility behaviour applies only

---

[1] RMB = renminbi, which is the name of the Chinese currency. At the current rate of exchange US$1 = RMB 8.3 approximately.

to married women aged 16–61 (born after 1961). For this particular group, there are altogether 1,665 observations in the sample, with 814 in Dongguan and 853 in Meizhou .

The data allow us to analyse the fertility behaviour of three groups of women, namely, non-migrants, permanent or *de jure* migrants, and temporary or *de facto* migrants, using a regression model. The estimated equations permit us to gauge to what extent fertility behaviour is related to the setting (for example, the position of the place of residence on the urban-rural continuum) and to a host of socio-economic variables indexing the level of economic development. Both the actual number of births and the desired number of births are analysed. The independent variables include:

(1) *Migration status according to household registration.* This is indexed by two dummy variables, the first being *de jure* migrant and the second *de facto* migrant. The hidden or reference category refers to non-migrant. The inclusion of this variable in the present study is self-explanatory.

(2) *Economic status according to household registration.* 1 = non-agricultural status; 0 = agricultural status. This variable is included to take into account the less strict application of the one child policy on the latter group (Siu, 1995).

(3) *Age.* The number of births is a positive function of age. This is because the maximum number of potential births corresponds to a woman's position over the span of child-bearing age. This is also because birth control measures in China have been gradually introduced and tightened over time.

(4) *Age at marriage.* Late marriage delays births. In fact, this constitutes an important part of the birth planning measures (Siu, 1995)

(5) *Position of the place of residence on the urban-rural continuum.* This is indexed by two dummies, the first is city (i.e., the former county seat) and the second is town. The hidden category is the rural areas. There may be milieu effects on fertility behaviour. In particular, it may be hypothesized that the rural setting tends toward more births and the urban setting toward more birth control practices.

(6) *Education.* This is given by three dummies: post secondary or above, senior secondary, and junior secondary. The hidden category

is primary or below. Education affects preferences, including preferences for children. A negative relation between education and the number of births can be expected.

(7) *Occupation.* The following occupational categories are employed: farmer, ordinary worker, petty trader, cadre and professional worker, others and unemployed. The hidden category is farmer. In China the work unit plays a significant role in effecting social control, including birth control (Siu, 1995). Farmers and petty traders fall outside the work unit system and are thus subject to somewhat less surveillance by the upper authorities. Cadres and professional workers are expected to act as models for the ordinary people. More stringent birth control measures may thus apply. However, it is the cadres who control the resources and therefore they are in a better position to evade birth control regulations.

(8) *Household Income.* Income is an important indicator of economic development. In this sense one would expect a negative relation between income and fertility. But households with higher incomes are more capable financially of having more children; at least they have the financial resources to pay the fines for excess births. Thus, the effect of household income on the number of births is indeterminate *a priori*.

Tables 1a and 1b list the variables (including the dependent variable) employed. Also included in the tables are the variables' respective means and standard deviations, both for the entire sample and for each of the Dongguan and Meizhou samples. In terms of the actual number of births, the mean for Dongguan (2.37) is essentially the same as that for Meizhou (2.35). The desired number of births, as can be expected, is somewhat higher than the actual number. The mean for the entire sample is 2.71. The difference between the Dongguan (mean = 2.81) and Meizhou (2.60) samples is not large, but it is statistically significant. Despite the inclusion of a supplementary sample of temporary migrants in the case of Dongguan, the percentage of married women in the temporary or *de facto* migrant category (5.15 percent) for Dongguan is not substantially higher than that for Meizhou (3.00 percent). On the other hand, the Meizhou sample has substantially more permanent or *de jure* migrants than the Dongguan sample. As for economic status as defined by household registration, the difference between the two samples is not too large. In both samples, slightly more

Table 1a. Mean values and standard deviations of the variables used in the regression analysis

	Entire sample		Dongguan		Meizhou	
	Mean	S.D.[1]	Mean	S.D.	Mean	S.D.
Actual no. of births	2.3596	1.2665	2.3707	1.2566	2.3493	1.2763
Independent variables:						
Migratory status						
de jure migrant	0.1286	0.3348	0.0965	0.2955	0.1585	0.3654
de facto migrant	0.0403	0.1969	0.0515	0.2211	0.0300	0.1707
Non-migrant	0.8311	0.3751	0.8520	0.3555	0.8111	0.3917
Age (10 years)	3.9029	1.0287	3.9157	1.0065	3.8910	1.0495
Age at marriage (10 years)	2.2075	0.3690	2.3094	0.3733	2.1124	0.3385
Urban-rural continuum						
City	0.2093	0.4069	0.2368	0.4254	0.1837	0.3875
Town	0.1267	0.3327	0.2008	0.4008	0.0576	0.2332
Rural	0.6640	0.4727	0.5624	0.4965	0.7587	0.4285
Hukou economic status						
Agricultural	0.5317	0.4991	0.5006	0.5003	0.5606	0.4966
Non-agricultural	0.4683	0.4992	0.4994	0.5003	0.4394	0.4966
Education						
Post-secondary or above	0.0230	0.1499	0.0180	0.1331	0.0276	0.1640
Senior secondary	0.2006	0.4006	0.1171	0.3218	0.2785	0.4485
Junior secondary	0.3994	0.4899	0.3320	0.4713	0.4622	0.4989
Primary or below	0.3770	0.4848	0.5329	0.4993	0.2317	0.4225
Occupation						
Ordinary worker	0.2559	0.4365	0.3243	0.4684	0.1921	0.3942
Peasant	0.3503	0.4772	0.1996	0.4001	0.4910	0.5002
Petty trader	0.0571	0.2322	0.0811	0.2731	0.0348	0.1834
Cadre or professional worker	0.1075	0.3098	0.0849	0.2790	0.1285	0.3348
Others	0.0472	0.2121	0.0553	0.2288	0.0396	0.1952
Unemployed	0.1820	0.3860	0.2548	0.4360	0.1140	0.3181
Household income (RMB 1000)	0.8176	1.9087	1.0836	2.7014	0.5695	0.3342
NO. OF CASES	1610		777		833	

[1] S.D. stands for standard deviation.
Source: Survey data.

Table 1b. Mean values and standard deviations of the variables used in the regression analysis

	Entire sample		Dongguan		Meizhou	
	Mean	S.D.[1]	Mean	S.D.	Mean	S.D.
Desired no. of births	2.7100	1.0500	2.8100	1.0800	2.6000	1.0200
Independent variables:						
Migratory status						
de jure migrant	0.1297	0.3361	0.0972	0.2964	0.1608	0.3676
de facto migrant	0.0462	0.2101	0.0627	0.2426	0.0305	0.1721
Non-migrant	0.8241	0.3808	0.8401	0.3660	0.8087	0.3939
Age (10 years)	3.8608	1.0425	3.8686	1.0221	3.8533	1.0621
Age at marriage (10 years)	2.2095	0.3709	2.3132	0.3695	2.1107	0.3444
Urban-rural continuum						
City	0.2120	0.4089	0.2386	0.4265	0.1866	0.3898
Town	0.1249	0.3307	0.1968	0.3978	0.0563	0.2307
Rural	0.6631	0.4731	0.5646	0.4963	0.7571	0.4295
Hukou economic status						
Agricultural	0.5315	0.4991	0.5018	0.5003	0.5599	0.4967
Non-agricultural	0.4685	0.4992	0.4982	0.5003	0.4401	0.4967
Education						
Post-secondary or above	0.0240	0.1532	0.0185	0.1347	0.0293	0.1689
Senior secondary	0.2006	0.4006	0.1193	0.3244	0.2782	0.4484
Junior secondary	0.4036	0.4908	0.3395	0.4738	0.4648	0.4991
Primary or below	0.3718	0.4835	0.5227	0.4998	0.2277	0.4199
Occupation						
Ordinary worker	0.2577	0.4375	0.3272	0.4695	0.1913	0.3936
Peasant	0.3488	0.4769	0.1993	0.4001	0.4917	0.5002
Petty trader	0.0595	0.2366	0.0824	0.2752	0.0376	0.1902
Cadre or professional worker	0.1051	0.3068	0.0836	0.2770	0.1256	0.3316
Others	0.0487	0.2152	0.0566	0.2312	0.0411	0.1986
Unemployed	0.1802	0.3845	0.2509	0.4338	0.1127	0.3164
Household income (RMB 1000)	0.8210	1.8797	1.0797	2.6438	0.5741	0.3385
NO. OF CASES[2]	1665		813		852	

[1] S.D. stands for standard deviation.

[2] The number of cases in Table 1a is slightly different from that in Table 1b because of missing values.

Source: Survey data.

than half of the married women are in the agricultural category. The two samples also differ very little in terms of the mean age of the respondents, which stands at about 39 years. However, the Meizhou respondents, in comparison with the Dongguan respondents, tend to marry at a younger age. The mean ages at marriage for the Dongguan and Meizhou samples are, respectively, 23.1 and 21.1 years. The urban-rural mix of the sample is, to a significant extent, predetermined by construction. But if we restrict ourselves to the target group, then from Table 1 it is clear that proportionately more of the sampled women in Dongguan, as compared with those in Meizhou, reside in the former county seat and towns. Earlier it was pointed out that people in Meizhou generally are better educated than people in Dongguan. This is also the case for the married women in the sample. Some 30 percent of the sampled women in Meizhou have senior secondary or above education; however, the corresponding figure for Dongguan is only 14 percent. In terms of occupation mix, the two samples are also quite different. Proportionately there are more farmers and fewer ordinary workers in Meizhou than in Dongguan, which is in line with the two cities' different positions on the ladder of economic development. However, Meizhou, in comparison with Dongguan, also has more married women who are cadres or professional workers, and fewer in the not employed (mainly engaged in homemaking) category. This is likely a reflection of Meizhou's status as the political centre of the Kejia people. In line with the difference in their level of economic development, the mean monthly household income for the Dongguan sample (at RMB 1084) is substantially higher than that for the Meizhou sample (at RMB 570).

Results of the regression analysis

Tables 2 and 3 give the regression results for the actual number of births and desired number of births, respectively. In each table the coefficient estimates of three regression equations are given, namely, the equation for the entire sample and the equations for Dongguan and Meizhou, respectively, when these two samples are analysed separately. In both the actual birth and desired birth equations, an analysis of covariance shows that partitioning the sample according to the place of the survey is statistically significant.[2] That is, the regression coefficients for the independent variables

---

[2] See Appendix I for the results of the covariance analysis.

Table 2. Results of the regression analysis: Actual number of births

Independent variable	Entire sample	Dongguan	Meizhou
Migratory status			
de jure migrant	0.016	0.018	0.086
de facto migrant	−0.128	−0.330**	0.165
Age (10 years)	0.876**	0.877**	0.901**
Age at marriage (10 years)	−0.702**	−0.724**	−0.802**
Urban-rural continuum			
City	−0.179**	−0.415**	−0.045
Town	−0.058	−0.221*	−0.03
Hukou economic status			
Non-agricultural	−0.383**	−0.264**	−0.340**
Education			
Post-secondary or above	−0.315*	−0.060	−0.334
Senior secondary	−0.203**	−0.076	−0.084
Junior secondary	−0.084	−0.038	0.016
Occupation			
Ordinary worker	−0.099	−0.197*	−0.214
Petty trader	0.104	0.022	−0.058
Cadre or professional worker	−0.179*	−0.415**	−0.212
Others	−0.273**	−0.391**	−0.327*
Unemployed	0.130	0.044	−0.007
Household income (RMB 1000)	0.017	0.011	0.097
Constant	0.813**	1.017**	0.733**
N	1610	777	833
R^2	0.641	0.661	0.637
F value	177.446**	92.546**	89.661**

* Significant at $p = 0.05$.
** Significant at $p = 0.01$.
Source: Survey data.

vary with the place of the survey. Separate regression equations are therefore needed for Dongguan and Meizhou. All the regression equations are significant at the .001 level. The R^2 obtained for the actual birth equations ranges from .63 to .66, and the R^2 for the desired birth equations ranges from .41 to .46. The explanatory powers obtained, especially the ones for the actual birth equations, are quite respectable for a regression using

Table 3. Results of the regression analysis: Desired number of births

Independent variable	Entire sample	Dongguan	Meizhou
Migratory status			
de jure migrant	−0.077	−0.135	0.086
de facto migrant	−0.360**	−0.564**	−0.189
Age (10 years)	0.494**	0.461**	0.563**
Age at marriage (10 years)	−0.302**	−0.394**	−0.382**
Urban-rural continuum			
City	−0.235**	−0.556**	.0.039
Town	−0.159*	−0.423**	0.112
Hukou economic status			
Non-agricultural	−0.480**	−0.365**	−0.252*
Education			
Post-secondary or above	−0.505**	−0.446	−0.279
Senior secondary	−0.360**	−0.255*	−0.104
Junior secondary	−0.169**	−0.095	0.002
Occupation			
Ordinary worker	0.098	−0.110	−0.257*
Petty trader	0.367**	0.276*	−0.173
Cadre or professional worker	0.104	−0.184	−0.228
Others	0.090	−0.059	−0.180
Unemployed	0.198**	0.008	−0.098
Household income (RMB 1000)	0.024*	0.017	0.174*
Constant	1.823**	2.471**	1.368**
N	1665	813	852
R^2	0.408	0.459	0.414
F value	71.000**	42.199**	36.862**

*   Significant at $p = 0.05$.
** Significant at $p = 0.01$.
Source: Survey data.

micro-level data. Higher explanatory power means less randomness. The result indicates that the actual conduct of fertility behaviour is subject to rather restrictive constraints, which bring about the relative lack of randomness of actual births. To facilitate our discussion, we first focus on the set of equations pertaining to actual births. We then discuss the results for the desired birth equations.

A. Actual number of births

To a certain extent the results of the regression analysis are in line with our expectations. In both Dongguan and Meizhou, a woman's economic status as defined by household registration is intimately related to her fertility behaviour. Specifically, a non-agricultural status (versus an agricultural status), on average, results in approximately 0.3 fewer births in both places. Also, in both Dongguan and Meizhou age is positively related to fertility but age at marriage is negatively related to fertility. The magnitudes of their effects in the two cities are also very much the same. For age, on average every ten years older in age is associated with about 0.9 more births. For age at marriage, deferring marriage by one year will result in 0.07 to 0.08 fewer births.

Another variable which exhibits similar effects (or, rather, in this case the lack of effects) in both cities is household income. In both instances, the coefficient estimate takes on a positive value. That is, household income is positively related to the number of births, after controlling for education, occupation and other socio-economic variables. To the extent that income is an indicator of economic development, this result contradicts the hypothesis that development is the best contraceptive (Siu, 2000; Teitelbaum, 1975). But this relation is not significant statistically. Also, development means much more than an increase in income.

Education, which influences the preference system, also fails to emerge as a strong factor affecting fertility. It is only marginally significant for the Meizhou equation and it is not significant for the Dongguan equation, although in both cases the corresponding coefficient estimates have the correct sign. Perhaps the lack of significance for education is due to the presence of multicollinearity, as education is closely related to other independent variables, such as the urban-rural continuum, age, occupation and income. There is evidence to support this interpretation. First, in the equation pertaining to the whole sample (hence with a much larger number of observations, which mitigates multicollinearity), both the post-secondary or above and senior secondary dummies are significant. Second, if we perform a regression exercise using only the education dummies, both the resulting equation and the individual dummy variables are highly significant (results not shown). These results hold for both the Dongguan and Meizhou samples. But the lack of significance for education in the two equations given in Table 2 also suggests a possible gap between preference and actual behaviour, as the latter is distorted by the enforcement of birth control measures.

Occupation is a purely categorical variable. Five occupation dummies enter the regression equations. The direction of the influence and the magnitude of the coefficient estimates for these variables in the two cities are very much the same. But the Dongguan equation shows stronger statistical significance. In this equation three of the dummy variables, namely ordinary worker, cadre and professional worker, and "other employed", are significant. However, in the Meizhou equation, only the "other employed" dummy is significant. Again, it could be that in Dongguan the birth control measures are applied differentially to different occupation groups, whereas in Meizhou they are not. This point will be further elaborated upon below when we discuss the desired birth equations.

From the above it is clear that the Dongguan and Meizhou equations bear many similarities. But the two estimated equations also contain certain major and interesting differences. Consider the migrant status variable. While in both cities no significant difference in the number of births is found between the permanent migrants and the non-migrants, in Dongguan the temporary migrant dummy is highly significant but in Meizhou it is not. To complicate the situation further, the coefficient estimate for the Dongguan equation is of the "wrong" sign. The result shows that married women in the temporary migrant category, on average, give 0.33 fewer births than the non-migrants, after controlling for a host of socio-economic variables. This indicates that the alleged adverse effects of the increasing incidence of temporary migration on birth planning are likely to be no more than impressionistic statements that are not grounded on concrete empirical evidence. It is true that it is difficult to monitor the birth planning programme of people who have migrated without a proper change in their household registration. Also, there is the problem of lack of clarity as to where the responsibility should lie. But the temporary migrants are faced with all sorts of uncertainties. They are discriminated against in the job market — the local authorities at their destination only care to find jobs for their "own" people (Li, 1997). The jobs taken by the temporary migrants are often those lower status and insecure jobs that the local people are not willing to take. Incomes are therefore low and unstable. Accommodations are also a problem. A substantial proportion of the temporary migrants live in factory dormitories or at construction sites, so they are not living with their families, including their spouses. Others have to rent tiny flats or cubicles from the local people, and they are therefore liable to eviction in the case they can no longer afford to pay the rent. Having children in the family would only aggravate the housing problem. Household income may be cut as a result as well. This is because the wife has to stay home to take care of the children. Worst still, as temporary migrants, the children are not entitled to compulsory education provided by the local authorities. Thus, the opportunity costs for having children are particularly high for the temporary migrants. All these factors argue for fewer births. Therefore, for the temporary migrants, there are two sets of forces, diametrically opposed to each other, acting upon fertility behaviour. In the case of Meizhou where migration is more related to family reasons, the two sets of forces seem to have more or less balanced each other out. However, in the case of Dongguan where migration is primarily economically driven, the latter set of forces dominates. The net result is a reduction in the number of births.

The position of the "place of residence on the urban-rural continuum" is another variable that has rather dissimilar effects in these two cities. In Dongguan, this variable is highly significant, and the magnitude and direction of its influence is in line with expectation: women in cities and towns, on average, give .42 and .22 fewer births, respectively, than women in rural areas. In Meizhou, however, the coefficient estimates for both the city and town dummies are not significant. Perhaps in Dongguan birth planning measures are more stringently enforced in the urban settlements, in line with state policy, whereas in Meizhou they are stringently enforced also in the rural areas because of the city's special political status. But this difference between the Dongguan and Meizhou results could also arise due to a difference in the preference structure between the two places. This point will be taken up in our discussion of the desired number of births.

B. Desired number of births

The results for the desired number of births are basically the same as those for the actual number of births. Again, the *de facto* migrant dummy for Dongguan carries a negative and highly significant coefficient estimate. However, for the Meizhou equation this variable is not significant. In a sense the consistency between the two sets of equation, i.e., desired and actual births, with respect to migrant status is to be expected, as the actual fertility behaviour of the temporary migrants is less distorted by state policy because of their transient nature.

As in the case of the actual birth equations, in both the Dongguan and Meizhou desired birth equations, the coefficient estimate for age is positive and significant. However, that for age at marriage is negative and significant. Also, in both equations the non-agricultural status dummy carries a negative sign and is significant. The results show that these variables not only affect actual fertility behaviour, as they are related to the implementation of birth control measures, but they also have a significant influence on people's preference for children. Thus, the younger women, who, in comparison with the older generation, tend to marry at a later age, are less resistant to birth planning. Also, with urbanization gathering pace, the household economic status of a growing proportion of the population will be converted from agricultural to non-agricultural. This again will have a dampening effect on fertility. These considerations suggest that the crude birth rate is unlikely to show a major rebound even if some of the birth control measures are relaxed. But birth planning is nevertheless important, as a comparison of the magnitudes of the coefficient estimates of the actual and desired birth equations shows that the effects of these variables on actual births are much larger than their effects on desired births.

The results for the urban-rural continuum dummies are also very much the same as those in the previous table. In the Dongguan equation, as before, both the city and town dummies are significant and carry a negative sign. In the Meizhou equation, also as before, both dummies are insignificant. Thus, in Dongguan women in cities and towns have less preference for children. However, in Meizhou such differences in fertility preference between urban and rural areas is not discerned. A plausible explanation for the differential effect of this variable in the two places may be that in the more prosperous Dongguan, the market plays an increasing role in structuring all aspects of social and economic behaviour and begins to have an effect on the preference system, including the preference for children. But its effect is mainly felt in the urban areas. In the rural areas the traditions still prevail. However, in less economically developed Meizhou, market and economic considerations play only minor roles, even in the urban areas. Traditions prevail in all types of settlements, whether they be urban or rural.

Education affects a person's world view. However, as in the case of the actual birth equations, in general the education dummies for the desired birth equations are not significant, although they have the expected sign and respective order of magnitude. Again, this is likely due to multicollinearity. Entering the education dummies as the only set of regressors yields highly significant estimates for both Dongguan and Meizhou, explaining some 12–18 percent of the sample variance. This perhaps explains why the mean number of desired births in Dongguan is larger than that in Meizhou, as a much larger proportion of the population in Meizhou is in the senior secondary education or above category.

Desired births, of course, are not the same as actual births. The latter are subject to the birth control measures and, as a result, they are generally fewer. The discussion above shows that the effects of variables such as younger age and non-agricultural status on reducing the number of actual births are larger than their effects on reducing the number of desired births. The two sets of equations also differ from each other in some other ways. It was pointed out above that in the Dongguan actual birth equation, both the ordinary worker and cadre and professional worker dummies are significant. The result shows, on average, compared to farmers ordinary workers have .20 fewer children, and cadres and professional workers .42 fewer children. But these two dummies fail to show significance in the desired birth equation, after controlling for other demographic and socio-economic factors. Of course, there may be collinearity problems. But it may also be argued that these people have preferences similar to those of the farmers: the history of rapid economic growth and urbanization is too short to have substantially changed their preferences. It is only because of the tighter controls in the work place of these occupation groups that there have been significantly fewer births.

Household income is another variable that varies somewhat between actual and desired births. In the case of the former, both the Dongguan and Meizhou regression coefficients are insignificant. In the case of the latter, while the Dongguan coefficient is still insignificant, the Meizhou coefficient is positive and significant at the .05 level. Higher incomes encourage a desire for more births. But the birth control policy prevents this from becoming a reality.

Concluding remarks

The data presented above show that the relation between regional development, migration and fertility is highly complex. Despite the two cities' rather different cultural traditions and economic performance in the reform era, the mean actual number of births per married woman in the

Dongguan and Meizhou samples is essentially the same. In fact, the Dongguan sample gives a higher mean desired number of births than the Meizhou sample. There is thus an apparent lack of association between level of economic development and fertility behaviour. Economic growth seems to have encouraged more births. An increase in fertility is only kept at bay by stringent application of birth planning.

But the above is only a casual interpretation of the findings. The regression results reveal that fertility is related to a host of social and economic variables. For example, in Dongguan residence in the former county seat (city) and the market towns, which is associated with the level of urbanization, results in substantially fewer births, although this relation is not evident in Meizhou. Being in the non-agricultural status category brings about similar effects. Engaging in urban-related jobs (versus working as farmers), especially in Dongguan, also results in fewer actual births, although its effect on desired births is insignificant. Education fails to emerge as a significant variable for either the actual birth or desired birth equations. But further analysis of the data suggests that this may only be due to multicollinearity, and education is likely to have a strong negative effect on fertility, whether it is measured in the actual or desired number of births. Higher household income, on the other hand, seems to encourage a greater desire for births, but this has not been translated into more actual births. All these argue that economic development does have an effect on fertility, although not necessarily a negative one. The almost identical mean number of actual births between the two samples is probably a mere coincidence, resulting from their rather different demographic compositions and from some of the same factors acting differently upon fertility behaviour in the two cities. The result of the covariance analysis demonstrates that the partitioning of the data according to the place of the survey is meaningful in a statistical sense. Consideration of place specificity is important in the study of fertility behaviour in China.

In contradistinction to popular belief, the findings also show that temporary migration is associated with fewer births, although this result only holds for the Dongguan equation. The dampening effect arising from the uncertainties in connection with temporary migration apparently overshadows the enhancing effect due to a relaxation of surveillance by local authorities. This is especially the case if the migration is driven by economic motivations, which is generally true for the Dongguan migrants. While there may be cases in which a woman migrates in order to escape from birth planning, these cases apparently do not dominate the migration scene. The Chinese authorities do not have to be overly concerned about them.

References

Brown, Lester. *Who Will Feed China: Wake-up Call for a Small Planet*. London: Earthscan Publications Ltd., 1995.

Chan, Kam Wing, Ta Liu and Yunyan Yang. "Hukou and non-hukou migrations in China: comparisons and contrasts." *International Journal of Population Geography*, 5 (1999): pp. 425–448.

Fan, Cindy C. "Economic opportunities and internal migration: a case study of Guangdong Province, China." *The Professional Geographer*, 48.1 (1996): pp. 28–45.

Gui, Shixun. *Zhongguo Liudong Renkou Jihua Shengyu Guanli Yanjiu (A Study of the Management of Birth Planning for the Floating Population)*. Shanghai: East China Normal University Press, 1992.

Ketizu (Study Group). *Zhongguo Tudi Ziyuan Shengchan Nengli Ji Renkou Chengzailiang Yanjiu (A Study of the Productive Capacity of Land Resource and Population Carrying Capacity of China)*. Beijing: People's University Press, 1992.

Li, Si-ming. "Population migration, regional economic growth and income determination: a comparative study of Dongguan and Meizhou, China." *Urban Studies*, 34 .7 (1997): pp. 999–1026.

Li, Si-ming and Yat-ming Siu. "A comparative study of permanent and temporary migration in China: the case of Dongguan and Meizhou, Guangdong Province." *International Journal of Population Geography*, 3 (1997): pp. 63–82.

Li, Si-ming and Yat-ming Siu. "Population Mobility" In *Guangdong: Survey of a Province Undergoing Rapid Change*, edited by Y. M. Yeung and David K. Y. Chu, 2nd edition, pp. 405–434. Hong Kong: The Chinese University Press, 1998.

Siu, Yat-ming. "*Cong Shunjing dao kunjing: jiushi niandai Zhongguo jihua shengyu de wenti* (From Smooth to Troubled Waters: Problems Concerning China's Birth Planning in the Nineties)." In *Zhongguo Shehui Fazhan: Xianggang Xuezhe de Fenxi (China's Social Development: Analysis of Hong Kong Scholars)*, edited by Si-ming Li, Yat-ming Siu and Tai-kei Mok, pp. 170–193. Hong Kong: Hong Kong Educational Books Ltd., 1995.

Siu, Yat-ming. "Is development the best contraceptive? An examination of the relationship between economic development and fertility in two counties in

Guangdong." In *China's Regions, Polity and Economy: A Study of Spatial Transformation in the Post-Reform Era*, edited by Si-ming Li and Wing-shing Tang, pp. 391–402. Hong Kong: The Chinese University Press, 2000.

Siu, Yat-ming and Si-ming Li. "Population Mobility in China." In *China Review 1993*, edited by Joseph Y S Cheng and Maurice Brosseau, pp. 19.1–19.31. Hong Kong: The Chinese University Press, 1993.

State Statistical Bureau of China (SSB). *Quanguo 1% Renkou Chouyang Diaocha Ziliao (Tabulations on the National 1% Population Sample Survey)*. Beijing: China Statistics Press, 1997a.

State Statistical Bureau of China (SSB). *Quanguo 1% Renkou Chouyang Diaocha Ziliao: Guangdong Fence (Tabulations on the National 1% Population Sample Survey: Volume on Guangdong)*. Beijing: China Statistics Press, 1997b.

State Statistical Bureau of China (SSB). *China Statistical Yearbook*. Beijing: China Statistics Press, 1999.

Statistical Bureau of Guangdong (SBG). *Guangdong Statistical Yearbook*. Beijing: China Statistics Press, 1999.

Teitelbaum, M. S. "Relevance of demographic transition theory for developing countries." *Science*, 188 (1975): pp. 420–425.

Wei, Bian. "200 million less births over 20 years." *Beijing Review*, Nov. 30–Dec. 6 (1998): p. 21.

Yeh, Anthony G. O. and Xia Li. "Economic development and agricultural land loss in the Pearl River Delta, China." *Habitat International*, 23.3 (1999): pp. 373–390.

Zhou, Yujun. "*Shichang jingji tiaojianxia liudong renkou dui jihua shengyu de zhenfu xiaoying* (Positive and Negative Effects of the Floating Population on Birth Planning under Conditions of a Market Economy)." *Nanfang Renkou (Southern Population)*, 2 (1996): pp. 22–26.

Appendix I
Dongguan and Meizhou: Covariance analysis on number of birth equations

A. Actual number of births

	SS explained	SS residual	DF
Entire sample	1653.192	927.584	1593
Dongguan (DG)	809.676	415.575	760
Meizhou (MZ)	863.931	491.411	816
DG+MZ	1673.607	906.986	1576
Incremental SS	20.415		16
Mean square	1.275938	0.575499	
F	2.217099		

At p = .05 and 16 and 1576 degrees of freedom, the critical value for F is 2.01.

B. Desired number of births

	SS explained	SS residual	DF
Entire sample	755.279	1095.692	1648
Dongguan (DG)	432.756	510.198	796
Meizhou (MZ)	368.289	521.414	835
DG+MZ	801.045	1031.612	1631
Incremental SS	45.766		16
Mean square	2.860375	0.632503	
F	4.522312		

At p = .05 and 16 and 1631 degrees of freedom, the critical value for F is 2.01.

4

Permanent Migrants, Temporary Migrants, and the Labour Market in Chinese Cities

C. Cindy Fan

Department of Geography, University of California, Los Angeles

Introduction

Migration and the development of a labour market are among the most revealing facets of the rapid transformation of Chinese cities during the past two decades. Both processes are intricately related to the household registration (*hukou*) system, one of the most definitive legacies of China's socialist institutions. At the same time, structural changes in the form of transition from a central planning mode toward marketization, from an economy dominated by the state sector to one with a more heterogeneous mix of activities and ownership types, and from urban economies guided by socialist models to cities with increasingly diverse and commercialized functions define the contexts for understanding the interrelationships among migration, labour market processes, and the *hukou* system.

Since the late 1950s, the *hukou* system has been used to monitor and control population movements, especially those to urban areas, in China

Acknowledgements: This research was funded by grants from the National Science Foundation (SBR-9618500), the Luce Foundation and the UCLA Academic Senate. I wish to acknowledge the contributions by Kam Wing Chan, Ling Li and Yunyan Yang to a collaborative project from which part of the data for this paper are drawn.

(e.g., Shen and Tong, 1992). *Hukou* is a record of one's (1) registration classification, and (2) registration location, and it is usually passed from one generation to the next. Registration classification refers to the "nonagricultural" and "agricultural" categories, designated respectively to the urban population entitled to state benefits and subsidies, and to the rural population that receives little state support other than the right to farm. Registration location refers to where a person's *hukou* "resides," which essentially records where he/she belongs. Having one's *hukou* in a city, or obtaining a local *hukou* in cities, is an entitlement to fully enter the labour market and to have access to subsidized benefits such as housing and education. It is very difficult for peasants to obtain nonagricultural *hukou* and to "move" their *hukou* to urban areas, making their survival in cities difficult. Therefore, for decades the *hukou* system has tied Chinese peasants to the countryside and has contributed to low levels of urbanization and mobility (Hsu, 1994; Wong and Huen, 1998).

But the socialist transitional process is marked by constant adjustments of state agencies and institutions to new demands and circumstances (Solinger, 1999). The magnitude of surplus agricultural labour, the infusion of foreign investment seeking cheap labour, and the marketization and commercialization of an urban economy that demands labour in industrial and services work have all prompted the state to create new channels for peasants to migrate to work in towns and cities. A variety of new innovations such as "temporary residence permits" and "identification cards" have become available since the mid-1980s to facilitate migration (e.g., Chan and Zhang, 1999).[1] However, by denying local urban *hukou* to peasant migrants, the state retains its role of gate-keeper of urban permanent residence. Most peasant migrants living and working in cities continue to hold agricultural *hukou* and their *hukou* still reside in their home villages. Their migration is considered "self-initiated" rather than sponsored by the state, and it is not accompanied by a *hukou* transfer to the cities. They are referred to as "temporary migrants." Without a local *hukou*, temporary migrants are excluded from many jobs (especially in the state sector) and from subsidized benefits.

At the same time, migrants who move to jobs assigned by the state, marry migrants to rural areas, and migrants who enter universities are eligible to obtain local *hukou* in their destinations. They are referred to as "permanent migrants." Though the state's roles in job assignments (to school graduates) and job transfers of state employees have subsided in recent years, they continue to be a factor of labour allocation that is responsible for the slow development of a labour market in China (Knight and Song, 1995; Maurer-Fazio, 1995). Not only do permanent migrants in cities have the legitimacy and right to stay there, they also have access to an array of jobs closed to temporary migrants, including high-paying and secure positions that are accompanied by health care and other benefits. On the other hand, temporary migrants in essence are considered "outsiders" and they are expected eventually to return to their origins. Many take jobs that are shunned by local residents. In the view of the state, permanent migration is official and "within the state plan," whereas temporary migration is unofficial and "outside of the state plan." In other words, the migrants' residence status, namely their *hukou*, symbolizes their geographical (rural versus urban) origins, defines their opportunities and constraints, and connotes their socioeconomic status (Cheng and Selden, 1994; Christiansen, 1990). In the city, a dichotomy of access, opportunities, and socioeconomic status is immediately associated with one's residence. A variety of terminologies have been used to address this dichotomy — permanent versus temporary migration, *hukou* versus non-*hukou* migration, "plan" versus "non-plan" or self-initiated migration, and formal versus informal migration (Chan, 1996; Chan et al., 1999; Fan, 1999; 2001; Goldstein and Goldstein, 1991; Goldstein and Guo, 1992; Yang, 1994; Yang and Guo, 1996). This paper adopts the terms permanent migration and temporary migration because they are most widely used in the literature.

Increased volumes of rural-urban migration and the centrality of *hukou* to migrants' opportunities and constraints have important implications for the evolution and organization of the Chinese urban labour market. The development of a labour market in Chinese cities has been accelerated by their transformation from "producing" entities where social and residential stratifications were minimized, heavy industry was emphasized, the tertiary sector was kept small, and city size was strictly controlled, to "consuming" entities characterized by consumerism, burgeoning markets, divisions of labour, a thriving service sector, a growing middle class, and a more international and Western outlook (French and Hamilton, 1979; Lo, 1994; Lo et al., 1977; Yang and Guo, 1996; Wang, 2000). Services such as domestic work, hotels and restaurants, repair shops and hair salons, and industrial work in foreign-invested enterprises have expanded and accelerated social stratification in cities. These jobs are at the lower end of the occupational hierarchy, and are mostly filled by temporary migrants.

The complexities of the migration system, labour market development and the continued roles of socialist institutions (especially *hukou*) in China call for analytical approaches that can cut across all three dimensions. The labour market segmentation theory offers a logical starting point for such efforts. It is especially relevant because of the emphasis on the contrasts between a primary or formal sector and a secondary or informal sector, and on the relationship between rural-urban migration and the segmentation process. The most popular representation of the theory involves the bifurcation of a labour market into a primary sector with relatively stable, high-skilled jobs offering high pay and good benefits, and a secondary sector with less stable, low-skilled, low-paying jobs with few benefits (e.g., Piore, 1979). This theory is especially popular among studies on developing countries that are characterized by a dual economy model with a Harris-Todaro (1970) type of rural-urban migration (Gupta, 1993). Migrants who cannot be absorbed by the primary sector in cities have accelerated the growth of a secondary sector (e.g., McGee, 1982). But most studies assume that migrants to cities are relatively homogeneous, and very few researchers examine the heterogeneity among migrants and how that affects the segmentation of the labour market.[2] In this paper, I argue that *hukou* and the distinction between permanent migration and temporary migration are central to an understanding of migration and labour market processes in Chinese cities.

The interrelationships among *hukou*, migration and labour market development are poorly understood and are mostly examined based on broad perspectives and macro-level data. This paper uses a combination of census-type data and survey data, and aims at documenting salient differences between migrants with different residence statuses and describing their varied migration and labour market processes. Specifically, I employ data from the 1990 Census and from a survey conducted in Guangzhou in 1998. Guangzhou is one of the biggest cities in China, its economy is diverse, and it has attracted a large number of migrants from rural and urban areas from throughout the country. In order to evaluate the role of residence status on migration and labour market processes, the analysis in this paper is guided by a comparative approach that seeks to identify the most salient differences between permanent migrants and temporary migrants. In addition, comparisons between these two types of migrants and non-migrants are also made, wherever possible. Though non-migrants and permanent migrants both have local *hukou*, for the sake of convenience I refer to these three sub-populations — non-migrants, permanent migrants and temporary migrants — as three types of residence status.

The 1990 Census and the 1998 Guangzhou survey

The empirical analysis uses data from a one-percent sample of the 1990 Census[3] and a survey in Guangzhou conducted in 1998. The 1990 Census defines a migrant as an individual five years or older whose usual place of residence on July 1, 1985 was in a different city, town, or county than that on July 1, 1990, and (1) whose *hukou* was in the 1990 place of residence (permanent migrants); or (2) who had stayed in the destination for more than one year or had left the *hukou* location for more than one year (temporary migrants). By definition, the 1990 Census underestimated the actual volume of migration because it excluded moves within cities or counties, migrants younger than five years old, migrants who died between 1985 and 1990, multiple moves between the two years, return migrants, and migrants who did not satisfy the "more than one year" requirement (mostly temporary migrants). But the 1990 Census inquired about the reason for migration, which makes it a more useful source of migration data than more recent national-level surveys such as the 1995 One-Percent Population Sample Survey.

Typically, census-type data provide information about general demographic patterns, while survey-type data can yield more in-depth information about specific processes. In June and July of 1998, I conducted a questionnaire survey in Guangzhou,[4] in order to examine migration and labour market processes in more detail. The survey (1998 Guangzhou survey) included three types of respondents — 305 non-migrants, 300 permanent migrants, and 911 temporary migrants. A larger number of temporary migrants was included because they are the newest and most dynamic migrants in Chinese cities and because their migration and labour market processes are less well understood. The sample was arrived at using stratified quota sampling, with stratification both across major occupational categories and geographic districts in Guangzhou. The Appendix describes the sampling framework and the sampling process.

In the Guangzhou survey, I employ definitions for non-migrants, permanent migrants and temporary migrants that are somewhat different from those in the 1990 Census. Specifically, non-migrants refer to indivi-duals who lived in Guangzhou for at least 15 years and whose *hukous* were in Guangzhou; permanent migrants refer to migrants who moved to

Guangzhou from 1990 and whose *hukous* were in Guangzhou; and temporary migrants refer to migrants who stayed in Guangzhou for at least three months but whose *hukous* were not in Guangzhou. I assume that individuals who moved to Guangzhou more than 15 years ago and who had local *hukou* were for all practical purposes non-migrants or permanent residents in that city. I use the "since 1990" criterion for permanent migrants in order to focus on relatively recent migrants whose moves were accompanied by a *hukou* transfer or who had obtained a local *hukou* in Guangzhou by the time of the survey. A "three months" rather than the census "one year" criterion for temporary migrants was used because the latter would likely exclude large proportions of temporary migrants in large cities such as Guangzhou. Finally, the survey includes only individuals ages 15 or older. Although the "duration of stay" and "arrival time" criteria are quantitatively different from those in the 1990 Census, the qualitative definitional distinctions, between temporary migrants and permanent migrants, and between migrants and non-migrants, are similar.

The Guangzhou survey has two limitations that must be taken into consideration before interpreting the data. Firstly, a survey of one site, albeit a major city, should not be considered representative of other urban areas in China. Secondly, unlike the 1990 Census, the volume, occupational and gender distributions of the 1998 Guangzhou survey sample are a function of the sampling framework and should not be the objects of inquiry (see Appendix). Despite these limitations, Guangzhou is a major magnet of migrants of all kinds; it has attracted large numbers of migrants from other parts of the province and from other provinces; it has a large, diverse and changing economy, and it has a rapidly expanding and relatively developed urban labour market. For these reasons, Guangzhou is among the best field sites to study migration and labour market processes in Chinese cities.

Demographic characteristics

Comparisons between the 1998 Guangzhou survey sample and the 1990 Census can help us assess the quality of the survey data and reveal broad differences between non-migrants, permanent migrants and temporary migrants. The two samples selected from the 1990 Census for comparison are the Guangzhou (*shiqu* or urban districts) sample and the cities (all *shiqu* or urban districts) sample, respectively, based on a one-percent sample of the 1990 Census (see note 3) (Table 1).

Table 1. Comparison of the 1998 Guangzhou survey sample with the 1990 Census samples: demographic characteristics

| | 1998 Guangzhou survey* | | | 1990 Census** | | | | | |
| | | | | Guangzhou (ages 15+) | | | Cities (ages 15+) | | |
	Non-migrants	Permanent migrants	Temporary migrants	Non-migrants	Permanent migrants	Temporary migrants	Non-migrants	Permanent migrants	Temporary migrants
N	305	300	911	31,822	4,038	2,567	1,986,736	97,929	85,477
%				82.8	10.5	6.7	91.5	4.5	3.9
Age									
Mean	35.9	29.8	27.5	39.5	23.2	28.4	37.6	27.1	29.9
15–39 (%)	65.2	89.7	90.2	56.8	95.4	86.7	61.5	86.9	82.6
Sex ratio	107	142	128	112	427	186	105	137	141
Nonagricultural hukou (%)	92.1	97.7	32.4	82.2	99.3	19.6	52.9	89.1	17.1
Place of birth (urban) (%)	90.8	69.4	29.9	—	—	—	—	—	—
Education (%)									
Illiterate/Primary	8.9	2.7	17.3	30.8	2.3	23.2	41.7	12.1	34.2
Junior high	26.2	5.3	53.3	30.2	5.1	51.1	34.8	22.5	49.1
Senior high	35.7	13.7	19.1	25.2	12.9	14.6	15.0	16.0	11.5
Above senior high	29.2	78.3	10.2	13.8	79.7	11.1	8.5	49.4	5.2
Occupation (%)									
Professional	17.2	46.0	7.2	25.4	61.0	12.4	18.4	39.7	6.4
Commerce	26.0	27.9	26.5	7.6	2.5	11.2	5.3	4.9	12.9
Services	18.2	16.1	27.9	9.1	4.7	13.5	5.2	5.8	12.4
Industrial	32.8	8.1	33.9	40.9	29.6	60.4	30.8	34.1	57.3
Agriculture	5.7	2.0	4.5	16.9	2.2	2.6	40.3	15.5	10.9

* Source: 1998 Guangzhou survey.

** Source: 1990 Census one-percent sample.

While the sample size (N) of the 1998 Guangzhou survey is a function of research design, those of the 1990 Census samples can be used to estimate population size. In 1990, a total of 17.2 percent (10.5 percent permanent migrants and 6.7 percent temporary migrants) of the Guangzhou population ages 15 and above were migrants, compared with 8.3 percent (4.5 percent permanent migrants and 3.9 percent temporary migrants) in all Chinese cities. The higher rate of in-migration in Guangzhou reflects the attractiveness of Guangdong as a migration destination as well as the high level of mobility within the province (Fan, 1996). In both the 1990 Census samples, permanent migrants accounted for larger proportions than temporary migrants. But the Census definitions that required a "more than one year" residence, described earlier, likely underestimated the number of temporary migrants in cities.

In all three samples, non-migrants were the oldest, as demonstrated by their higher mean ages and smaller proportions in the 15–39 age group, whereas both permanent and temporary migrants were heavily concentrated in the 15–39 age group. In the 1990 Census samples, temporary migrants were older than permanent migrants, but in the 1998 Guangzhou survey sample permanent migrants (29.8) were older than temporary migrants (27.5). Migrants had higher sex ratios than non-migrants, which is consistent with the conventional wisdom that men are more mobile than women. But the sex ratio of permanent migrants in the 1990 Census Guangzhou sample was exceedingly high — 427. These age and sex ratio discrepancies are probably related to educational attainment, occupational distributions and reasons for migration, which will be elaborated upon below.

As expected, in all three samples, the vast majority of non-migrants and permanent migrants, and only a minority of temporary migrants, held nonagricultural *hukou*. The relatively low proportion of non-migrants in the 1990 Census cities sample holding nonagricultural *hukou* (52.9 percent) suggests that some cities included significant proportions of the rural population within their administrative (*shiqu*) boundaries. This observation is further supported by the relatively high proportion of non-migrants in the cities sample who engaged in agriculture (40.3 percent). But Guangzhou was more urbanized than the average city in China, hence a high proportion of non-migrants holding nonagricultural *hukou* (82.2 percent) and a low proportion of them engaging in agriculture (16.9 percent) in 1990. The place-of-birth data from the 1998 Guangzhou survey, which show that 69.4 percent and 29.9 percent of permanent migrants and temporary migrants respectively were born in urban places (cities and towns), further highlight the contrasts in background between them — permanent migrants were more likely to have urban backgrounds and temporary migrants were more likely to have rural backgrounds.

In all three samples, permanent migrants were the most highly educated, with "above senior high" educations, while temporary migrants were concentrated at the "junior high" level. Non-migrants were less concentrated at specific education levels, but in general they were more highly educated than temporary migrants. As a whole, the 1998 Guangzhou survey sample and the 1990 Census Guangzhou sample were more highly educated than the 1990 Census cities sample, which reflects Guangzhou's relatively high level of development as well as its concentration of educational institutions. A comparison of the 1998 Guangzhou survey sample with the 1990 Census Guangzhou sample suggests that the former resembles the educational attainment of permanent migrants and temporary migrants in that city. Non-migrants in the 1998 Guangzhou survey sample, however, had higher education levels than their counterparts in the 1990 Census Guangzhou sample. All in all, the 1998 Guangzhou survey sample yields a clear educational ranking among the three types of residence status: permanent migrants, non-migrants, and temporary migrants in descending order.

In both the 1990 Census samples, permanent migrants were most highly represented in the professional (including administrative) occupations, while temporary migrants were most highly represented in industrial work. In the 1990 Census Guangzhou sample, in particular, professional occupations accounted for 61.0 percent of permanent migrants, and industrial work accounted for 60.4 percent of temporary migrants. These observations underscore a strong relationship between the *hukou*-based migration system and segmentation of the labour market. In both the 1990 Census samples, the economic structure of the respective cities was likely the main driving force of the non-migrants' occupational distributions. For example, Guangzhou's industrialized and relatively developed economy resulted in higher proportions of non-migrants in industrial (40.9 percent) and professional (25.4 percent) occupations than the average city in China.

As described earlier and in the Appendix, the occupational distribution of the 1998 Guangzhou survey sample is a result of a predetermined sampling framework. Specifically, using the 1990 Census Guangzhou sample as a reference, I increased the proportions in commerce and services for all three subpopulations, in order to adjust for the expansion of these occupations in Guangzhou during the 1990s. Correspondingly, the relative proportions of professional and industrial occupations were adjusted downward. The net result was that the three subpopulations surveyed had the same modes, with deflated proportions, as the 1990 Census Guangzhou sample, namely, industry (32.8 percent) for non-migrants, professional (46.0 percent) for permanent migrants, and industry (33.9 percent) for temporary migrants.

The concentration of permanent migrants in the "above senior high" education level and in the professional occupation in Guangzhou is probably related to their high sex ratio — 427 according to the 1990 Census — that was observed earlier. The concentration of educational institutions in Guangzhou has attracted large numbers of migrants who want to pursue higher education. This kind of migration is mostly represented as permanent migration, as admission to universities entitles the migrants to obtain a local *hukou* (see also Table 2). The large number of migrants moving to Guangzhou to pursue education has probably resulted in a relatively young average age among permanent migrants in that city, as observed earlier. But persistent gender discrepancies in access to higher education suggest that there are considerably more men than women who participate in migration for educational purposes, boosting the sex ratio among permanent migrants. Similarly, the higher representation of men than women in professional work also contributes to a high sex ratio among permanent migrants.

The above analysis highlights important socioeconomic differences among the three subpopulations, and underscores residence status as an important definition of social and economic stratification in Chinese cities. In general migrants are younger than non-migrants. But a young age is probably the only similarity between permanent and temporary migrants. Non-migrants and permanent migrants are associated with urban back-grounds, and temporary migrants are associated with rural backgrounds. Contrasts in terms of human capital are marked by a hierarchy in educational attainment, especially notable in Guangdong where permanent migrants are the most highly educated and temporary migrants are the least educated. These discrepancies in human capital and backgrounds have probably resulted in differentials in labour market opportunities, as illustrated by the observed hierarchy in occupational attainment, with permanent migrants again at the top and temporary migrants at the bottom.

Table 2. Comparison of the 1998 Guangzhou survey sample with the 1990 Census samples: migration

| | 1998 Guangzhou survey* | | 1990 Census** | | | |
| | | | Guangzhou (ages 15+) | | Cities (ages 15+) | |
	Per-manent migrants	Tem-porary migrants	Per-manent migrants	Tem-porary migrants	Per-manent migrants	Tem-porary migrants
Origin type (%)						
City	59.5	17.0	43.9	8.3	31.7	10.4
Town	22.4	25.3	33.4	14.8	25.4	13.5
Village	18.1	57.6	22.7	76.8	42.9	76.1
Interprovincial (%)	49.5	53.9	20.1	16.5	25.3	34.2
Source of interprovincial migrants (%)						
Eastern	37.2	16.8	33.0	11.3	38.2	47.2
Central	51.4	66.9	55.6	80.4	39.3	33.8
Western	11.5	16.2	11.4	8.3	22.5	19.0
Reason for migration (%)						
Job transfer	15.7	1.3	5.3	1.0	20.5	5.2
Job assignment	24.0	0.4	17.4	0.4	12.7	2.8
Industry/business***	14.3	93.1	0.9	66.5	1.7	57.8
Study/training	28.3	0.7	68.2	10.2	34.8	3.5
Friends/relatives	—	—	1.2	6.7	6.5	12.1
Retirement	—	—	0.2	0.5	1.3	0.8
Joining family	5.7	2.1	3.6	1.6	8.8	4.2
Marriage	7.3	1.3	1.0	4.1	7.6	9.0
Other	4.7	1.1	2.2	9.0	6.1	4.6

*   Source: 1998 Guangzhou survey.

** Source: 1990 Census one-percent sample.

*** Note: "Industry/business" in the 1990 Census is considered the same as "self-initiated" moves in" the 1998 Guangzhou survey.

Migration processes

In both the 1990 Census Guangzhou and cities samples, the majority of permanent migrants originated from villages, and the majority of permanent migrants originated from cities or towns (Table 2). The 1998 Guangzhou survey sample shares these general characteristics. Specifically, 57.6 percent of the temporary migrants surveyed came from villages, and 81.9 percent of the permanent migrants surveyed came from cities or towns. These statistics again indicate that permanent migrants are more likely to come from urban areas and that temporary migrants are more likely to come from rural areas.

But the 1998 Guangzhou survey sample seems to have over-represented inter-provincial migration. The respective proportions of inter-provincial migrants among permanent and temporary migrants — 49.5 percent and 53.9 percent — are significantly higher than those of the 1990 Census samples (see also Figure 1). This discrepancy likely is related to the higher representation of commerce and services, which tend to attract large numbers of inter-provincial migrants. But the regional origins of inter-provincial migrants in the 1998 Guangzhou survey sample resemble those in the 1990 Census Guangzhou sample. Specifically, the majority of the inter-provincial migrants originated from provinces in the central region, but significant proportions of the permanent migrants came from provinces in the eastern region. This pattern is consistent with the observation in recent research that eastward movements from inland to coastal provinces dominate inter-provincial migration, but it also suggests that permanent migrants are more likely than temporary migrants to have origins in the more developed eastern region.

The 1998 Guangzhou survey data permit more detailed analyses of the migrants' origins. As shown in Figure 1, the origins of the permanent and temporary migrants differed. Specifically, higher proportions of temporary migrants originated from other provinces and the periphery (northern and non-coastal parts) of Guangdong, and a higher proportion of permanent migrants originated from non-Pearl River Delta open areas in Guangdong. These differences suggest that permanent migrants are more likely to come from more developed origins and temporary migrants are more likely to come from less developed localities. But the higher proportion of temporary migrants originating from the Pearl River Delta (16.5 percent) seems to refute the above generalization. A closer examination of the data shows that 66.0 percent of the temporary migrants who originated in the Pearl River Delta were not first-time migrants (Figure 2). In other words, they had migrated from other provinces or other parts of Guangdong to the Pearl River Delta before their most recent migration to Guangzhou. In fact, among temporary migrants originating from Guangdong, the proportion of non-first-time migrants grew with an increasing level of development — lowest for the peripheral origins (16.9 percent) and highest for Pearl River Delta

Figure 1. Origins of migrants: 1998 Guangzhou survey.

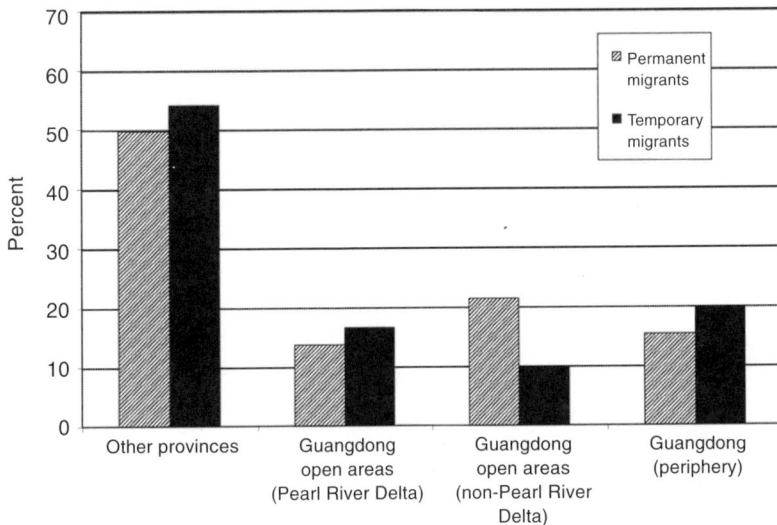

Figure 2. Non-first-time migrants: 1998 Guangzhou survey.

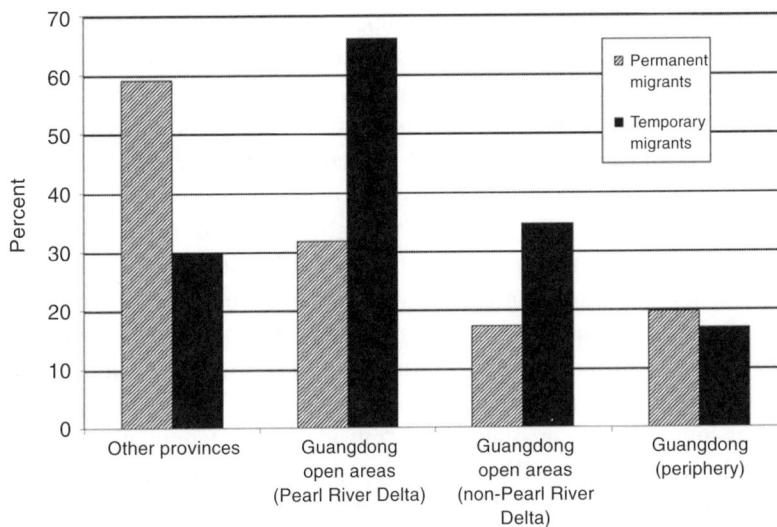

origins (66.0 percent) — suggesting a pattern of migration in stages from less developed areas to more developed areas.

Permanent migrants and temporary migrants differed considerably in their reason for migrating. Here, a clarification of "reason for migration" is in order. The standard nine "reasons for migration," listed in Table 2, have their origins in Chinese census-type surveys, including the 1987 One-Percent Sample Survey and the 1990 Census. But the definitions of these "reasons" indicate that they denote the means and types of migration and the degrees of state involvement more than the motives behind population movement (Fan, 1999).[5] For example, "job transfer" and "job assignment" refer to the state's allocation of human resources by assigning school graduates and transferring workers to specific jobs and regions; "industry/business" refers to self-initiated migrants wishing to engage in industrial, commercial or trade sectors; "friends/relatives" refers to migration to seek the help of friends and relatives, and it is mostly associated with self-initiated migration; and "study/training" refers to migration to attend schools or training programs. While "job transfer" and "job assignment" are results of the state's role in shaping the labour market, "industry/business" is most closely identified with migration "outside of the state plan."

In all three samples, "industry/business" was the dominant reason for temporary migration, while "study/training," "job assignment" and "job transfer" were the leading reasons for permanent migration. These differences highlight the varied opportunities available to migrants. Migrants whose jobs are allocated by the state and those who migrate for education purposes — "within the state plan" migrants — are eligible for local *hukou* in the cities, while "outside of the state plan" migrants who move on their own initiative and are not sponsored by the state by and large are not eligible for city *hukou*.

"Study/training" was the most prominent reason for permanent migration, accounting respectively for 34.8 percent, 68.2 percent and 28.3 percent of permanent migrants in the 1990 Census Cities sample, the 1990 Census Guangzhou sample, and the 1998 Guangzhou survey sample. These statistics reflect the concentration of educational institutions in Chinese cities. The prominence of "industry/business" among temporary migrants highlights the availability of "outside of the state plan" employment opportunities, especially in industry, services and commerce, in large cities such as Guangzhou. "Marriage" was relatively less important in all three samples, since most marriage migrants in China are rural-rural migrants and are not highly represented in the cities (Fan and Huang, 1998).

In general, the 1998 Guangzhou survey sample reflects the types of migration observed in the two 1990 Census samples as well as the adjustments made in the sampling framework. The larger proportion of temporary migrants selecting "industry/business" (93.1 percent), and the lower proportion of permanent migrants selecting "study/training" (28.3 percent), are partly due to the higher representation of commerce and service occupations in the sampling framework. But as a whole the distribution of reasons for migration in the 1998 Guangzhou survey sample closely resembles the contrasts observed in the two 1990 Census samples — permanent migration is more closely associated with institutional opportunities related to education and state allocation of human resources, while temporary migration is overwhelmingly represented by self-initiated moves.

While census-type "reasons for migration" focus on migration types and institutional involvement more than migration motives and decision-making, the 1990 Guangzhou survey data permit more detailed analyses of the latter (Table 3). "Job search" was the most important motive among both permanent and temporary migrants surveyed; and "increase income" and "study" were the second leading motives respectively for temporary migrants and permanent migrants. The greater prominence of "job search" and "increase income" reasons among temporary migrants suggests that direct monetary returns and improvements in economic well-being were the major forces behind their moves. Temporary migrants' reasons for leaving their origins — 63.5 percent citing "low income" and 19.1 percent "few jobs," compared respectively with 28.0 percent and 5.7 percent among their permanent migrant counterparts — further support this observation. These statistics, combined with earlier observations about migrant origins, suggest that temporary migrants to Guangzhou are mostly "upward" movers from poorer and less developed areas, while permanent migrants are represented by both "upward" and "level" movers, including those from urban and more developed areas.

Finally, the reasons for choosing Guangzhou as a destination reflect the perception that there are available economic opportunities in the city (Table 3). The survey permitted multiple responses to this question, which partly explains the scattered distribution of the responses. Nevertheless, "higher wages" and "ease in finding jobs" accounted respectively for

Table 3. Migration considerations: 1998 Guangzhou sample

	Permanent migrants	Temporary migrants
Motive for migration (%)		
Job search	38.5	55.7
Increase income	11.7	37.7
Family/marriage	15.1	3.2
Study	28.8	1.1
Other	6.0	2.2
Reason for leaving origin (%)		
Low income	28.0	63.5
Few jobs	5.7	19.1
Family	16.0	8.8
Study	39.3	0.6
Other	11.0	8.1
Reason for choosing Guangzhou (multiple) (%)		
Higher wages	30.8	23.2
Ease in finding jobs	19.0	32.1
Family/relatives	17.7	26.2
Proximity to origin	12.9	13.0
Other	19.6	5.5

55.3 percent of the responses by temporary migrants and 49.8 percent of those by permanent migrants. Among both permanent and temporary migrants, "proximity to origin" was not an important reason for choosing Guangzhou, which not only reflects the large proportion of interprovincial migrants among the survey sample, but also points to the strong economic pull of Guangzhou which likely offsets the friction of distance for migrants.

Labour market processes

The data on occupational distribution that were summarized earlier suggest that the labour market in Guangzhou and in other Chinese cities is highly segmented, and that the segmentation is correlated with residence status. The 1990 Census Guangzhou sample, in particular, shows that permanent migrants concentrated in professional occupations, temporary migrants concentrated in industrial work, and non-migrants were well represented in both (Table 1). In short, permanent migrants had the highest occupational levels, followed by non-migrants and finally temporary migrants.

A comparison of the job search experiences of non-migrants, permanent migrants and temporary migrants further reveals important processes relevant to labour market segmentation. While census-type data do not provide such information, the 1998 Guangzhou survey data can shed some light on job search experiences in that city. Table 4 shows that "income" was the leading job search criterion of all three subpopulations, and that it

Table 4. Job search and sector: 1998 Guangzhou sample

	Non-migrants	Permanent migrants	Temporary migrants
Criteria for job search (%)			
Income	40.9	45.3	62.7
Ownership sector	18.6	23.0	4.7
Stability	24.9	13.2	18.8
Location	1.7	7.4	3.6
Benefits	4.7	5.7	3.2
Other	5.3	4.1	3.7
Information about labor market (%)			
Relatives in Guangzhou	50.8	49.0	41.2
Relatives outside Guangzhou	0.7	4.7	36.9
Advertisement	11.5	15.3	8.8
Work unit/school	22.4	11.0	0.2
Agencies in Guangzhou	4.1	6.0	4.1
Agencies outside Guangzhou	0.0	0.3	1.7
Other	9.8	13.7	6.6
Medium for this job (%)			
Self	55.7	64.0	87.1
Recruitment	23.9	21.7	7.6
Work unit assignment	18.7	9.0	0.7
Other	1.6	5.3	4.6
Ownership sector (%)			
State	43.3	57.1	16.4
Collective	13.0	5.1	9.1
New-economy	22.0	20.9	55.0
Self-employed	21.7	16.9	19.4
Stability			
Number of jobs (mean)	2.2	1.9	2.5
Years at present job (mean)	9.9	3.9	2.7

was most prominent (62.7 percent) among temporary migrants. "Ownership sector" accounted respectively for 23.0 percent and 18.6 percent of non-migrants and permanent migrants, but only 4.7 percent of temporary migrants. These differences again suggest that income was a predominant incentive for temporary migration, while non-migrants and permanent migrants were driven not only by income but also by non-monetary considerations such as ownership sector (see below). "Location" and "benefits," on the other hand, were of little importance to all three subpopulations.

In the 1998 Guangzhou survey, respondents were asked to specify their main source of information about the labour market and the medium for finding their present job. Responses to the questions suggest that temporary migrants had less access to institutional resources than their non-migrant and permanent migrant counterparts, and that they had to rely on self-help and network-type resources. Table 4 shows that "relatives in Guangzhou" was the leading source of labour market information for all three types of samples. But "relatives outside Guangzhou" was also a very important source of information for temporary migrants (36.9 percent). Most likely, these were relatives from the migrants' home towns or villages. In other words, social and kinship networks were the dominant source of information for temporary migrants. Though networks were also important among non-migrants and permanent migrants, as indicated by the high percentages of "relatives in Guangzhou," the relative importance of "advertisements" and "work unit/school" suggests that these two groups were more connected to institutional and organized information sources. As a whole, the data show that a social network is generally important for a job search, but it is probably more important for temporary migrants who are relatively detached from institutions and other organized sources in Guangzhou.

As a result of differing occupational distributions, job search experiences and opportunities in the labour market, the sectoral distribution varies considerably among the three subpopulations. The analysis focuses on four ownership sectors — state, collective, new-economy and self-employed. The state and collective sectors are the traditional ownership types in the Chinese socialist economy. In large and older cities such as Guangzhou the state sector continues to be prominent. On the other hand, recent reforms of state-owned enterprises and changes in the urban economy have reduced the dominance of the state sector and promoted shifts of the labour force to the non-state sector. In order to address the complexity of the non-state sector, in the analysis it is further broken down into the "new-economy" and "self-employed" sectors. New-economy refers to employment in other than state- and collective-owned enterprises. This sector, mainly represented by foreign-invested, private, family and individual-owned enterprises, has rapidly gained prominence since the 1980s. It is characterized by jobs in industry and services, and is very important to temporary migrants who have less access to institutional resources and who have more opportunities outside the state sector. Secondly, the self-employed sector refers to employers and individuals who own their own businesses, as opposed to employees in state, collective or new-economy sectors. Like the new-economy sector, the self-employed sector has increased in size since the economic reforms. It is more developed in large and commercialized cities such as Guangzhou where reforms and a concentration of human resources have given rise to a more open and plural economy, to a critical mass of entrepreneurs, and to many business opportunities.

Data from the 1998 Guangzhou survey show that the state sector accounted for 57.1 percent of non-migrants, 43.3 percent of permanent migrants, but only 16.4 percent of temporary migrants (Table 4). The majority of temporary migrants — 55.0 percent — was in the new-economy sector. These statistics support the notion that non-migrants and permanent migrants have greater access to well-established and institutional labour market processes, while temporary migrants are mostly channeled to newer and less institutionalized segments of the urban economy. In addition, 16.9 percent, 21.7 percent and 19.4 percent respectively of non-migrants, permanent migrants and temporary migrants were in the self-employed sector, suggesting that self-employment has emerged as an important segment in Guangzhou's labour market.

Finally, Table 4 shows that job turnover was the highest among temporary migrants and lowest among non-migrants. The average number of jobs one has had was 2.2 for non-migrants, 1.9 for permanent migrants and 2.5 for temporary migrants. The relatively small number of jobs for non-migrants, despite their older ages, suggests a high level of job stability, which is further illustrated by their long duration at the present job (averaging 9.9 years). By contrast, temporary migrants' higher job frequency (2.5) and shorter duration at the present job (2.7 years) depict a relatively high level of job turnover, further underscoring their association with less established and less stable segments of the labour market.

Summary and conclusion

This paper has highlighted residence status as a critical determinant of the differences in migration and labour market processes in Chinese cities. Results from the empirical analysis suggest that the complex Chinese urban labour market is intricately related to migration processes and institutional legacies. Drawing from census data on Chinese cities as well as on a recent survey on Guangzhou, the findings support the argument that labour market segmentation in China must be understood and conceptualized in relation to residence status and the transition in the urban economy. Though the Guangzhou survey examines migrants and the labour market in only one city, that city's attractiveness to migrants and its diverse and relatively developed economy suggest that observations here may well be repeated in other large Chinese cities that have also received large numbers of migrants. In many ways, the findings from the survey support and complement those based on the 1990 Census.

There are clearly enormous socioeconomic, migration and labour market differences among non-migrants, permanent migrants and temporary migrants. The differences are especially acute between permanent and temporary migrants, and are in no small part due to a society-wide sorting mechanism that allocates urban *hukou* to migrants with higher educations from urban and more advantaged locations. They are, so to speak, state-sponsored migrants who are differentiated from those with low education from rural and poorer regions, who must migrate without the benefit of a local *hukou*. The sorting continues during migration and in the labour market, partly because of the migrants' different residence statuses and partly because of their human capital differences. Permanent migrants have greater access to institutional resources and information and are engaged in more prestigious occupations. By contrast, most temporary migrants are dislocated from institutional resources and are clustered in less prestigious occupations and sectors.

The findings in this paper support the notion that residence status is central to the processes of migration and segmentation of the urban labour market in China.[6] Attempts toward theorization of these processes must first take into consideration the selectivity, opportunities and constraints associated with residence status. In the city, the biggest divide appears to be between permanent migrants and temporary migrants, rather than between migrants and urban natives. In this regard, the traditional dual labour market model, which does not emphasize the heterogeneity among migrants or the institutional factors of migration, is useful but too simplistic for the Chinese case. The transitional setting, where socialist-type and newer economic activities are both in flux, and where market and institutional mechanisms coexist, adds further complexities to our understanding of migration and labour market processes in Chinese cities.

Appendix

A sampling framework with stratification across both occupational categories and the eight urban districts in Guangzhou was used to guide the sampling process. Using the distribution of major occupations in Guangzhou from the 1990 Census as a basis (see Table 1), four types of adjustments were made. The first adjustment was made to reflect the changes in the city's economic structure between 1990 and 1998, by increasing the relative proportions of commerce and services. I then adjusted the occupational proportions of the three types of residence status in order to estimate their likely occupational distributions. The actual occupational distributions of these three subpopulations in Guangzhou are not known because official data do not include migrants who have not registered with the local authorities, but who account for a significant proportion of the temporary migrants in the city. Therefore, I relied on informants in Guangzhou, including State Statistical Bureau survey specialists, as well as a variety of scholarly and journalistic sources that estimated the breakdowns of occupational categories (e.g., management, street vendors, garment workers, etc.) among migrants in Guangzhou. Thirdly, in order to facilitate comparison among the three subpopulations, adjustments were made to ensure that sufficient numbers in the respective occupational categories were included. For example, I adjusted upward the proportions of permanent migrants and non-migrants in commerce. Finally, I roughly allocated the proportions of men and women, except for the occupational categories that are clearly dominated by one sex (e.g., nannies and construction workers).

The geographic proportions for the initial sampling framework across the eight urban districts of Guangzhou were derived from existing data on the geographic distributions of population, nonagricultural population, and migrants in Guangzhou, as well as the settlement history of individual urban districts. For example, Yuexiu and Haizhu were allocated larger proportions because they are among the oldest and most urban parts of Guangzhou, and they are known for an increasing concentration of permanent and temporary migrants during the 1990s. Conversely, smaller proportions were allocated to surburban districts, such as Baiyun, that are less known for migrant concentration.

Using an initial sampling framework, a team of six interviewers employed the quota sampling technique and randomly interviewed respondents who satisfied the occupational and geographical criteria until the predetermined numbers or proportions in the initial sampling framework were reached. Table 1 shows the occupational breakdowns of the survey sample. See Fan (2001) for more details about the sampling and survey processes.

Notes

1. Since the early 1990s large cities have begun to offer "blue stamps" or "blue seals" (*lanyin*) *hukou* for sale, subject to regulations set by local governments. But most peasant migrants are not eligible for or cannot afford this relatively new type of *hukou* (Wong and Huen, 1998).
2. One exception is Gordon (1995), who differentiates contracted migration – migration secured with employment — from speculative moves where a job search follows migration and where employers seek to minimize their commitments and responsibilities toward labour.
3. The one-percent sample is a clustered sample containing information about every individual in all households of the sampled village-level units (villages, towns, or urban neighborhoods in cities) drawn from China's 1990 Census. It includes a total of 11,475,104 records.
4. The survey was designed together with Kam Wing Chan, Ling Li and Yunyan Yang.
5. See SSB (1993: 513–514, 558) and Fan (1999) for definitions of the nine reasons for migration.
6. See Fan (2001) for a discussion of the relationship between residence status and labour market returns.

References

Chan, Kam Wing. "Post-Mao China: a two-class urban society in the making." *International Journal of Urban and Regional Research,* 20.1 (1996), pp. 134–150.

Chan, Kam Wing, Liu, Ta and Yang, Yunyan. "*Hukou* and Non-*hukou* migrations in China: comparisons and contrasts." *International Journal of Population Geography*, 5 (1999), pp. 425–448.

Chan, Kam Wing and Zhang, Li. "The *hukou* system and rural-urban migration in China: processes and changes." *The China Quarterly*, 160 (1999), pp. 818–855.

Cheng, Tiejun and Selden, Mark. "The origins and social consequences of China's *hukou* system." *The China Quarterly*, 139 (1994) , pp. 644–668.

Christiansen, Flemming. "Social division and peasant mobility in Mainland China: the implications of hu-k'ou system." *Issues and Studies*, 26.4 (1990), pp. 78–91.

Fan, C. Cindy. "Economic opportunities and internal migration: a case study of Guangdong Province, China." *Professional Geographer*, 48.1 (1996), pp. 28–45.

———."Migration in a socialist transitional economy: heterogeneity, socioeconomic and spatial characteristics of migrants in China and Guangdong province." *International Migration Review*, 33.4 (1999), pp. 950–983.

———."Migration and labour market returns in urban China: results from a recent survey in Guangzhou." *Environment and Planning A* (2001), forthcoming.

Fan, C. Cindy and Huang, Youqin. "Waves of rural brides: female marriage migration in China." *Annals of the Association of American Geographers*, 88.2 (1998), pp. 227–251.

French, R. A. and Hamilton, F. E. I. "Is there a socialist city?" In *The Socialist City: Spatial Structure and Urban Policy*, edited by R. A. French and Hamilton, F. E. I., pp. 1–21. Chichester: John Wiley and Sons, 1979.

Goldstein, Alice and Guo, Shenyang. "Temporary migration in Shanghai and Beijing." *Studies in Comparative International Development*, 27.2 (1992), pp. 39–56.

Goldstein, Sidney and Goldstein, Alice. *Permanent and temporary migration differentials in China.* Honolulu: East-West Population Institute, Papers of the East-West Population Institute No. 117, 1991.

Gordon, Ian. "Migration in a segmented labour market." *Transactions, Institute of British Geographers*, NS 20 (1995) , pp. 139–155.

Gupta, Manash Ranjan. "Rural-urban migration, informal sector and development policies: a theoretical analysis." *Journal of Development Economics*, 41 (1993), pp. 137–151.

Harris, J. R. and Todaro, M. P. "Migration, unemployment and development: a theoretical analysis." *American Economic Review*, 60 (1970) , pp. 126–142.

Hsu, Mei Ling. "The expansion of the Chinese urban system, 1953–1990." *Urban Geography*, 15.6 (1994), pp. 514–536.

Knight, John and Song, Lina. "Towards a labour market in China." *Oxford Review of Economic Policy*, 11.4 (1995), pp. 97–117.

Lo, C. P., Pannell, C. W. and Welch, R. "Land use changes and city planning in Shenyang and Canton." *Geographical Review,* 67 (1977), pp. 268–283.

Lo, Chor Pang. "Economic reforms and socialist city structure: a case study of Guangzhou, China." *Urban Geography,* 15 (1994) , pp. 128–149.

Maurer-Fazio, Margaret. "Building a labour market in China." *Current History,* 94.593 (1995), pp. 285–289.

McGee, Terence G. "Labour markets, urban systems, and the urbanization process in Southeast Asian countries." *Papers of the East-West Population Institute,* 81 (1982) , pp. 1–28.

Piore, M.. *Birds of Passage: Migrant Labour in Industrial Societies.* New York: Cambridge University Press, 1979.

Shen, Yi Min and Tong, Cheng Zhu. *Zhongguo renkou qianyi (China's population migration).* Beijing: Zhongguo Tongji Chubanshe, 1992.

Solinger, Dorothy J. *Contesting Citizenship in Urban China: Peasant Migrants, the State, and the Logic of the Market.* Berkeley: University of California Press, 1999.

State Statistical Bureau (SSB). *Zhongguo 1990 nian renkou pucha ziliao (Tabulation on the 1990 population census of the People's Republic of China),* Vol. IV. Beijing: Zhongguo Tongji Chubanshe, 1993.

Wang, Feng. "Gendered migration and the migration of genders in contemporary China." In *Re-Drawing Boundaries: Work, Household, and Gender in China,* edited by Barbara Entwisle and Henderson, Gail. Berkeley: University of California Press, 2000.

Wong, Linda and Huen Wai-Po. "Reforming the household registration system: a preliminary glimpse of the blue chop household registration system in Shanghai and Shenzhen." *International Migration Review,* 32 (1998), pp. 974–994.

Yang, Quanhe and Guo, Fei. "Occupational attainment of rural to urban temporary economic migrants in China, 1985–1990." *International Migration Review,* 30.3 (1996) , pp. 771–787.

Yang, Xiushi and Guo, Fei. "Gender differences in determinants of temporary labour migration in China: a multilevel analysis." *International Migration Review,* 33.4 (1999), pp. 929–953.

Yang, Yunyan. *Zhongguo renkou qianyi yu fazhan de chanqi zhanlue (Long-term strategies of population migration and development in China).* Wuhan: Wuhan Chubanshe, 1994.

Urban Administration, Urban Development and Migrant Enclaves: The Case of Guangzhou

Wolfgang Taubmann

Department of Geography & Geology, University of Hong Kong;
Department of Geography, University of Bremen, Germany

Introduction

One of the most remarkable examples of the new dynamism accompanying China's economic restructuring and growth is the mass migration of peasants and rural inhabitants into the cities and the emergence of migrant settlements.

The new ruptures, as well as the tensions of a former almost egalitarian society, can mainly be seen in the outer areas of big cities — the zone where city and countryside meet. By taking migrant enclaves and mainly the "urban fringe" as examples, the processes of radical social change and the new management and planning problems of big cities are explored.

The municipality of Guangzhou is a very appropriate example, since the city was able to regain its former position as the southern gateway to China after the introduction of the economic reforms. This reestablishment of its importance as a consequence of the open-door policy can be documented by many indicators, such as the increasing gross value of both industrial output and exports or — most prominently — the growing role of the tertiary sector in Guangzhou's economy. Within the Pearl River Delta

I am indebted to Mr. Xu Xiaowei, Ph.D. candidate in the Department of Geography & Geology, Hong Kong University, who provided me with valuable material and established connections with institutions and persons in Guangzhou.

growth triangle Guangzhou is one of the major cities, ranking third after Beijing and Shanghai in terms of the number of the 500 largest foreign-funded industrial companies in China, import and export corporations and foreign financial institutes — to mention only a few examples (Zhou, 1999). Not only the influx of foreign investors, but also the emergence of private domestic capital and hundreds of thousands non-state enterprises were of vital importance to the growing economic role of Guangzhou. No wonder that the main strategy of party and government officials was and still is today to develop Guangzhou not only into an "economic centre in Guangdong province and southern China" but also into an "international metropolitan city, and the financial, trade and tourist centre in the Asia-Pacific region" (Xu, 1999, p. 121).

These ambitious development plans, combined with fast economic growth, have resulted in rapid urban restructuring and extension, driven by countless infrastructure and real estate development projects and an increasing consumption of living and urban space per resident.

The Comprehensive City Plan of 1994, the latest in a long row of such plans since 1949, though not yet officially approved by the State Council, should in principle be able to guide and control the scale, layout and pattern of Guangzhou's urban land uses. The prediction of the Guangzhou Urban Planning Bureau for the total amount of urban land use in 2010 is based on an estimated 4.08 million permanent urban residents and 1.5 million floating population. The assumed per capita urban land use is 90 square metres for permanent residents and 70 square metres for transient inhabitants (Xu, 1999, p. 133). Based on these assumptions, the total urban land use will reach about 555 square kilometres by 2010. Compared with 1980, when the built-up area covered only 136 square kilometres, this figure implies a quadrupling within thirty years. In the year 2000, the built-up area was supposed to reach already 335 square kilometres. Though the distinct stipulations of the 1989 City Planning Act should be able to guide the management of the land market and the changes in the use of land and buildings, the development of Guangzhou can be characterized as a "development-led" planning rather than a "planning-led" development (ibid., p. 256).

On any count, Guangzhou is one of the most rapidly expanding cities in southern China. The city actually has extended in nearly all directions, especially toward the north, southeast and southwest, though the original plan of the municipal government was to develop the city more toward the east. No doubt, the fast unplanned sprawl of residential and industrial development has been promoted at the expense of former agricultural land, especially in the north and northeast. In the late 1980s and early 1990s between 75 and 100 percent of land requisitioned by the Guangzhou municipal government for urban use was former agricultural land. In many cases, large-scale housing projects and economic and technical development zones are located at the urban fringe due to the acquisition of land from farmers (Yeh and Wu, 1996).

The expanding city, peasants and migrants

The massive transformation of agricultural land into urban and industrial uses is relevant to our topic, since many suburban peasants have lost their traditional livelihoods and have had to change to other economic activities. However, this transformation cannot yet be demonstrated by the official statistics, since the number of peasants in the urban district (*shi qu*) only decreased from 254,365 in 1989 to 239,456 in 1998 (Guangzhou Statistical Yearbooks, 1990 and 1999).

The other aspect, linked to the fast urban development, is the increase in the permanent and floating population. The number of permanent residents grew by nearly one million between 1988 and 1998 (1988: 5.77 million; 1998: 6.74 million) — almost exclusively limited to the population with an urban *hukou*, that increased from 3.26 million to 4.17 million (Guangzhou Statistical Yearbook, 1989 and 1999).

Any attempt to assess the floating/transient/temporary population in Guangzhou is rather uncertain. While the number stood at around 0.5 million in 1984, it was 1.17 million in 1988 and 1.3 million in 1989 (Chan, 1996; Chan and Gu, 1996). The actual figure varies considerably, depending on the source: According to the Public Security Bureau, 1.268 million temporary residents were registered in Guangzhou at the end of 1999, while the total floating population was estimated at 2 million (Guo, Li and Lin, 1999). Another source calculates the *wailai renkou* in Guangzhou's urban districts at 1.8 million (http:// dailynews7.sina.com.cn/2000-05-04/86999. html). The ratio between registered temporary residents and those without any registration is very difficult to assess. A sample survey carried out in 8 urban districts revealed that only 45.5 percent of the transient population had a proper registration. In Lujiang village/Haizhu district, only 50.5 percent of the transient population had applied for a certificate, according to a report by the Public Security Bureau (Research Department, 1999).

If we use the data given by the Public Security Bureau, we take rather conservative figures.

By 1999, most of the temporary residents were of working age and were part of the migrant labour force (94 percent). Of the workforce, most were employed as factory workers (69 percent), in the service sector (15 percent) or were self-employed (9 percent), worked as agricultural workers (3 percent) or were categorized as "others" (4 percent) (Guo, Li, and Lin 1999).

About 80 percent of all temporary residents are living in the urban fringe, mainly in the suburban district of Baiyun (28 percent), in the county-level city of Panyu (under the administration of Guangzhou) (21 percent), in the urban district of Haizhu (14 percent), in the county-level city of Zengchen (10 percent), and in the urban district of Tianhe (8 percent).

The distribution of the temporary population is closely correlated with the location of the so-called *chu zu wu*, i. e., apartments and houses for leasing mainly built by the farmers on their remaining collective land. This spatial coincidence of temporary residents and houses built by former farmers is a striking feature on the urban fringe of Guangzhou (see Table 1). The only exception seems to be in the district of Tianhe, where there is a higher share of *chu zu wu* compared to the percentage of temporary residents. This situation can be explained by a relatively large number of houses owned by permanent urban residents (4,397 out of 19,945 in 1998, according to the Public Security Bureau, Tianhe district).

Table 1. Distribution of "*chu zu wu*" and temporary residents in Guangzhou 1999

DISTRICT/CITY	CHU ZU WU (%)	TEMPORARY RESIDENTS (%)
Baiyun	31	28
Panyu	16	21
Haizhu	14	14
Tianhe	15	8
Others	24	29
Total	**100**	**100**
Abs. number	126,779	1,267,832

Source: Research Department, Household Management Department, PSB Guangzhou, 1999; Guo, Li and Lin, 1999.

Certainly not all migrants live in houses and apartments rented from peasants. There are other migrant living areas and living patterns in Guangzhou as well.

- A very small, but increasing number of migrant labourers lives in commodity housing provided by real estate companies. In these cases, the real estate companies are in charge of the proper management of the transient population and the necessary certificates.
- Migrants live in quarters owned by their employers or by their *danwei*. Here, the manager or boss is responsible for the proper registration. From the public security bureau's standpoint, this form of housing can be easily verified and therefore is a preferable mode of living.
- Migrants stay in hostels rented by enterprises from other partners, e.g., villagers' committees.
- By 1998, roughly 3,400 construction sites existed in Guangzhou, offering lodgings to nearly 209,000 migrant construction workers (Research Department, 1999).
- There are roughly 3,000 "underground" factories in Guangzhou — a settlement pattern that is regarded as being very disorderly. This type of housing and working is called "three in one", which means that working/producing, living and storing all take place in one place. These cramped quarters are mostly located at the urban fringe or in the near-urban countryside.
- Many governmental institutions at the central or provincial levels or from other provinces offer shelter and premises to migrant workers in order to earn extra income. These units are out of the control of the Guangzhou government and therefore present a major problem for the responsible authorities.

However, undoubtedly most migrants live in houses and apartments rented out by peasants. In 1999 about 46 percent of all registered temporary residents lived in such houses and apartments.

Of the 126,000 *chu zu wu* hired out, more than 90 percent were owned by farmers; the rest were owned by village committees and collective units or by some other local owners. A small part (5.9 percent) of the houses and apartments were rented to local permanent residents, to local companies or to some governmental institutions.

Table 2. "*Chu zu wu*" according to ownership and tenants

	1997	1998	1999
Total number of *chu zu wu*	98,396	102,779	126,142
Privately owned	—	93,158	113,834
Collectively owned	—	9,621	12,308
Tenant households	122,168	124,930	158,299
Temporary residents	428,074	461,832	585,00*

* Estimation based on the number of households.
Source: Guo, Li and Lin, 1999.

This development at the urban fringe has rarely been described in relevant publications dealing with Guangzhou, since the development process is incalculable, chaotic and driven by the profit-orientation of virtually tens of thousands of individual agents.

The boom of "*chu zu wu*" and urban development

In principle, there is a legal basis for the regulation of Guangzhou's suburban development. The 1989 City Planning Act provides the possibility for the establishment of development control plans based on administrative units. These plans are meant to control and guide land-use planning. For example, by 1998, 40 of the about 195 administrative villages within Baiyun district had carried out some kind of detailed construction plan. In 1999, another 61 villages were included, and in 2000 the remaining 94 villages completed such a local plan (Interview with cadres of the street bureau Kuangquan/ Baiyu District, March 2000; Study Group ... 1999).

Some of these villages are already completely surrounded by the expanding city, some are located at the edge of Guangzhou's built-up area and some more or less are still situated outside the city proper. In the first case, the former villages are relatively integrated into the urban area and little agricultural land remains, but even in the towns or villages outside the city proper some town governments have already secretly converted cultivated land into areas for commodity housing (Xu, 1999, p. 222). Furthermore, there is a brisk black market in the rural areas on the urban fringe, as farmers rent their land to other users because the price is higher than the compensation offered by the municipal government in regular transactions (Yeh and Wu, 1996, p. 349).

The so-called detailed plans to some extent might have prevented the present chaotic development concerning the conversion of agricultural land or the construction and development of the village enclaves. However, the main problem up to now is that the existing laws and regulations simply have not been enforced.

There is, for example, a wide variety of illegal construction activities, such as illicit allotments of building land and unlawful construction on allocated plots. Especially in possibly thousands of cases in the suburban districts of Baiyun and Fangcun commodity houses were built illegally on collectively-owned agricultural land by villagers, villagers' committees and township governments. Though all building activities require official approval, nearly everyone erects a building, mostly houses for lease, wherever she or he wants without consideration of the distance to the neighbouring property, to the streets or generally to the existing building regulations.

There are many plots where the ground area of a building covers nearly 100 percent of the site area. The clients often add one or more illegal floors or other structures to already approved buildings. In some cases house owners who originally received building permission for three to four storeys on a small plot actually build another six to seven storeys onto the existing house. For instance, a local self-employed person in Lujiang village/Haizhu district built an eight-storey house on a site of 200 square metres, containing 46 households and housing 136 tenants. In the same village a farmer demolished his four separate traditional buildings and erected a big yard for many migrants, managed by four full-time managers (Chen, 1999).

Some local officers describe these buildings in the village enclaves as "shaking hand buildings" since it is indeed easy to shake hands through a half-open window with a resident in an adjacent building and it is sometimes not possible to open a window fully due to the narrow lanes. Furthermore, in many cases the construction quality is very poor due to the lack of control.

The main differences between Guangzhou and other big cities such as Beijing or Shanghai regarding the building structure of the migrant enclaves seem to be that in the latter cities already existing farm houses or provisional structures are often used to house migrant workers, while in Guangzhou multi-storeyed buildings with rented flats are dominant. Therefore, the problems of crowding, inadequate infrastructure or insufficient lighting conditions seem to be even more severe in Guangzhou than in other cities.

Many villages still lack an urban infrastructure, such as a road network, electricity, water supply, drainage and sewage system, fire prevention lanes, basic urban services, and especially sanitary facilities. Often the hygiene situation is disastrous; streets do not have asphalt surfaces, and the risk of fire is high. Whereas the public facilities in many cases are totally neglected, the villagers seem to do whatever they can to maximise their housing areas within the allocated site, in many cases at public expense. Development in the urban-rural fringe is illustrative of the "tragedy of the commons" which primarily is attributed to capitalism. It means everyone pursues his own interests exclusively at the expense of public property, thus causing damage to the community.

In general, urban planning in Guangzhou for the migrant labourers' enclaves can at best be characterised as an "ad hoc" planning trying in retrospect to cope with the reality of the proliferation of these enclaves. In fact, there is almost no plan at all, or a local plan that only exists on paper.

Obviously, administrative competence seems to be a severe problem, as is the clash of interests between different levels of urban administrative units. The urban government often has inadequate personnel to deal with development and planning problems on the urban-rural fringe; on the other hand, these tasks are the responsibility of subordinate authorities at the district level. The urban planning law of Guangzhou requires street governments to support the district planning departments to control illegal construction; however the district authorities do not have the power to enforce the law and to punish clients who construct such illegal buildings. For example, the director of the responsible district department demanded that most of the illegal buildings erected in a village under the district's jurisdiction be demolished. However, the local village leader threatened him so as a result he did nothing. The head of a town under the jurisdiction of Baiyun district approved the construction of buildings on illegally allotted plots, even though the Guangzhou Municipal Government had issued a document stipulating that planning power delegated to local departments not be devolved to street bureau or township governments (Xu, 1999, p. 169). In the end, the town head not only was not punished but in fact was promoted. The internal comment given was, "laws are powerless against good relations" (*guanxi*).

Urban planning and development seem to bypass, or intentionally to overlook, the unmanageable enclaves that in reality have become "forgotten lands" at the edge of the city. Neither the city government nor the lowest level administrations are interested in carrying out serious planning in this zone. The local residents do not have the slightest knowledge about comprehensive urban planning goals and therefore build only according to their own interests. Future urban planning in this zone, if there will be any at all, will find it difficult to correct this disorderly development. The local governments and the local landlords clearly have the right to be kept in the know about urban planning; otherwise there is no incentive to act according to some comprehensive planning and development goals.

The relations between locals and outsiders

To illustrate the relationship between migrants and locals, the village of Lujiang/Haizhu district, which today is totally integrated into the built-up urban area of Guangzhou, will be examined (Interview with Prof. He and Mr. Sun, Department of Anthropology, Zhongshan University, March 2000).

Lujiang village, which had about 230 local households before 1949 (1, 100 persons), today houses 435 local households, and roughly 11,600 migrant workers. In the 1950s Lujiang was still a typical village near a large southern city of the pre-Communist era, which experienced the functional replacement of the kinship system by co-operative farming as the core of the village structure (Yang, 1959, p. 9). This traditional village is now part of the urban fringe and in a way it is an urban community. There are two interesting and even paradoxical developments. On the one hand, some traditional local features have been revived; on the other hand the economic activities have totally changed and the village is largely dependent on the transient population.

The traditional village as a unit of permanent community life was closed to outsiders. In the 1950s there were few families and individuals with surnames other than the five family clans. These persons were regarded as "floating elements", "sojourners", or "guests" whose life roots did not belong to this place (Yang, 1959).

While the number of outsiders was limited in the 1950s, today they are the overwhelming majority. However, the migrant workers of today are still regarded as outsiders and therefore are excluded from local society. But the economic relationship between outsiders and locals has changed dramatically. Today it is a kind of business-like relationship. The transient population is employed in township and village enterprises, rents flats or engages in independent businesses, including illegal activities such as prostitution, drug dealing etc.

While the relationship between the locals and the "outsiders" can be described as a modern economic trend, at the same time the local community is reviving some traditional activities, such as reconstruction of lineage temples, keeping of lineage books, and introducing traditional ceremonies. For instance, one of the clan temples was rebuilt in 1985 during a period of rapid urbanisation in this area. The revival of lineage connections during this rapid economic and social change created a kind of social security net for the local community. Some of the five traditional clans still exist.

As Zhou Daming (1997a) observes in a migrant enclave not far away from Lujiang village, migrant workers, particularly odd-job migrants, have developed subcultures of their own. They communicate primarily with people from their native regions, speak their native dialects, maintain their own eating habits, live in close communities, and have strong emotional bonds to their home districts. In Guangzhou such subcultures have not become integrated with the local culture, since migrant workers (from other provinces) do not speak Cantonese and do not have any contact with local people.

Many of the local families living within or adjacent to Lujiang or other village enclaves are now wealthy because of the windfall profits from rapid urbanisation and the exploitation of the migrant workers. In the first place, they are renting out houses and flats to the temporary residents. Second, the villagers' committees have erected village cooperative enterprises (*jingji hezuo qiye*) on collectively owned land or factory buildings that are leased to other enterprises or joint ventures. These collective economic activities are a second source of income for the local agricultural population. In Lujiang, each member of the local community is a shareholder in these activities and receives at least 16,000 RMB per year out of the distributed profits. In another village (Fenghua) the distributed profits amounts to 160,000 yuan annually. The 2,000 local residents in Yaotai village under the street bureau of Kuangquan/Baiyun district receive collective earnings of about 22,000 to 24,000 RMB annually.

The income from renting out property can easily amount to several thousand RMB per month. A unit of 50 square metres could be rented out for around 1,200 RMB per month during the middle of the 1990s. Since then, the demand has decreased and the rent accordingly; however, even today a monthly rent of 600 RMB is common for such a unit.

Additionally, most local families received high compensation for farmland that is leased or sold to governmental or private agencies and used for non-agricultural urban or industrial purposes.

Many local families have at least one to two members who still keep their agricultural *hukou*. This means that nearly every family benefits from collective economic activities. These benefits in some richer suburban rural areas have led some people who earlier had settled down in the cities to now ask to return, even going so far as to buy a rural *hukou* (Zhou Daming, 1997a, pp. 13ff.)

This new development has had a paradoxical effect: While on the one side traditional forms of village life have reappeared in the course of the economic reforms in a big city setting; on the other side new "modern" phenomena are developing. Since many peasant families have become rich, though lacking both working skills or flexibility to adapt to the new urban lifestyle, many of the younger local residents belonging to such families dislike working as unskilled and low paid workers or pursuing their educations. This phenomenon is called the transition from "peasant to playboy". Many of the younger villagers spend their time on illegal activities, such as gambling, drug dealing, pimping, and accepting and trading stolen goods (e. g. bicycles). In other words, a new criminal or semi-criminal relationship has developed between some locals and groups in the temporary population. According to a figure disclosed by the government, there are several thousand persons addicted to drugs in Guangzhou's urban fringe, the majority of whom belong to the "playboy generation" of villagers.

Since many permanent urban residents — and many administrative units — regard the floating people as the root of social unrest, it is not surprising that many labels are attached to the migrant enclaves. Living quarters dominated by migrant workers are often described as "paradise of thieves and robbers", "camps for prostitutes", "retreats of hunted criminals", "typhoon harbour for floating people escaping governmental control over family planning", or "underground factories for faked or inferior products". According to the Public Security Bureau, there was a significant increase in crime, such as homicide, robbery, bodily injury, and rape, between 1996 and 1998. Even new forms of organised crime and crime connected to gangs developed during this period (Research Department, 1999).

Though these designations in many cases may only mirror the usual prejudices of the local population, they also indicate a critical social situation and convey a feeling of insecurity. According to Solinger (1999, p. 131), a mid-1990s opinion poll of residents in Beijing, Guangzhou, and Shanghai found that the lack of social order was the "number one public enemy", as respondents regarded the floating population as the "root cause" of their feeling of danger. Presumably, this attitude is typical mainly among those local residents who have few personal contacts with migrants. Suburban villagers are more sympathetic to migrant workers, understanding how hard it was for them to leave their homes (Zhou Daming, 1997b, pp. 227ff).

Management and control of local house owners and the transient population

One of the main concerns of Guangzhou's administrative units at all levels is not the control of the floating population, but control of the house owners and landlords. In most migrant enclaves nearly all houses are rented out to the transient population while the house owners live elsewhere — a fact which makes control difficult. For example, in Yaotei village under the administration of Kuangquan street bureau 2,153 out of 2,450 private flats were for lease; 90 per cent of the flats and houses were owned by local villagers. Here and in other villages nearly all the houses were rented out to the transient population while the house owners were living elsewhere.

Therefore, the supervision of house owners and leases by migrant labour offices or police stations and the prevention of illegal leasing are regarded as highly necessary. As a rule, a landlord receives a renewable permission valid for one year to rent out his/her house or flat. In the eyes of the authorities it is essential that the landlords obey the rules. However, owners always tend to underreport the number of leased houses and apartments. Tax evasion is a common phenomenon because of the many undisclosed cases of leasing. House owners try to avoid reporting their income even if they have to pay a fine, which is up to 300 percent of the monthly rent. However, since there is no bookkeeping it is nearly impossible to calculate the income or profit. In many cases, the village leaders or the villagers' committees are quite unwilling to control the illegal activities since they themselves also rent their properties.

In general, house owners are out of control, and their activities are mainly determined by the pursuit of profit and not by any legal consciousness. Even if they report their tenants to the authorities, the official tenants and the actual occupants are often different persons. Furthermore, at the urban fringe illegal renting of welfare flats is quite common. If two persons in one household are allocated two different welfare apartments by their respective *danwei*, they often rent out one apartment to migrant workers.

In Baiyun district more than 36,000 flats and houses were registered in 1999, a number which is much below the real figure, since there is a large number of undisclosed cases (possibly the majority of all private houses in the urban-rural fringe — according to some sources). These houses — primarily located in the near-urban towns and villages — were rented by roughly 300,000 registered temporary residents (not including a large number of the out-of-town floating people who were staying without a temporary right of residence).

Since many houses were rented out without official permission, public control of safety precautions regarding construction or hygienic conditions is extremely difficult. In many cases urban administrative measures in the migrant enclaves have simply ceased. For example, an office of the Guangzhou Land and Housing Management Bureau only exists at the district level. Therefore, at the urban-rural fringe no one is responsible for the control of inadequate housing, unapproved buildings, infrastructure, environmental regulations, rights of ownership or use, or relations between landlord and tenant.

This lack of management and control is not due to a weak legal base, since actually there are many regulations and laws concerning landlord-tenant relations, e.g., "Regulations on rental of houses in cities and towns of Guangdong (issued by the People's Congress in Guangdong), "Management of renting of houses in Guangzhou" (issued in December 1995 by the city government) or "Some regulations concerning the rental of houses in Guangzhou's rural areas" (issued by the Guangzhou Land and Housing Management Bureau).

From the viewpoint of lower level administrative units, similar shortcomings also apply to control of migrant workers. Though there are controls in the form of periodical checks by officials of the village committee, many problems are overlooked. The street bureau relies on village committees or resident committees, but the village committees especially fulfil their duties rather poorly.

In principle, migrants have to provide three different certificates or documents — the so-called three documents (*san zheng*): a) a temporary residence certificate, if a person stays more than three days (this certificate might be valid for one month, three months, six months, or one year according to the migrant's own claim), b) a family planning certificate, c) an "outward working" permit, issued by the government of one's home town.

Guaranteeing universal control of the migrant people with respect to their documents seems to be an impossible task. The most simple indicators for this situation are — as mentioned above — the widely varying figures about the proportion of unregistered — and uncontrolled — migrants. Solinger (1999, p. 139) claims that up to 70 percent of the labouring migrants in Guangzhou were not registered.

The various problems linked to the management of the floating population can be illustrated by the example of Baiyun district.

According to information given by local officials, many problems are due to the parallel rural and urban administration systems — at least at the lowest administrative levels on the edge of the city of Guangzhou. As a rule, a village committee responsible for those inhabitants with a rural *hukou* should be looked after by the town government, while the residents' committee in charge of urban *hukou* inhabitants should be directly supervised by the street bureau. However, since some of the members of the traditional villagers' households hold a rural *hukou* and some an urban *hukou*, it is — from the administration's standpoint — difficult to manage and enforce any effective control over the residents. This situation causes low efficiency and results in administrative gaps.

In order to create a more effective base for social and security tasks, in May 1997 the district government enacted and promulgated "Measures for the administration of the floating population". Based on this document, a "Baiyun district leading group for the management of the floating (non-local) population" was set up, following a provincial regulation issued in 1993. The group members were mainly composed of district officials representing the majority of the government agencies in charge of the floating population, such as the housing management bureau, the public security bureau, the family planning committee, the urban construction office, the civil affairs bureau or the judicial authorities. A permanent office of the above-mentioned district leading group was also set up and attached to the public security bureau.

Similar corresponding groups were established at the town and street bureau level, however, under a different name ("leading group for the management of temporary residents"). At the lowest level of the administrative echelon — villagers' or residents' committees — so-called "managing groups for the floating population" were established.

According to a government document of May 1997, per every 80 households in public housing (state or collectively owned) and per every

50 households in privately owned houses (mainly houses for lease owned by villagers) a full-time or part-time member of the managing group, a so-called "assisting *hukou* manager" was appointed. Until the present, in Baiyun district there have been 276 managing groups and 2,162 assisting *hukou* managers (of whom 734 were full-time and 1,428 were part-time). This means the actual situation in Baiyun district, with a ratio of one assistant manager to roughly 180 households, is still far from the planned number.

In the entire city of Guangzhou in 1999 the figure stood at 2,226 household managing groups and 11,792 managing assistants (3,039 full-time and 8,353 part-time). Though the ratio of 1 to 135 is somewhat better than that in Baiyun district, it still does not reach the intended target.

The ratio between the number of assistants and the number of migrant households seems not to be the real problem. More serious is the low qualifications of the household managing assistants who have very rude manners and treat the migrant labourers poorly. It is not uncommon that migrants are beaten up by the assistants if they cannot pay the various fees and taxes. This is because the household management assistants are

Table 3. Management system for the floating population in Baiyun district/ Guangzhou at different levels:

District	"Baiyun district leading group for the management of the floating (non-local) population"	Various offices represented by officials: housing management bureau, public security bureau, family planning committee, urban construction office, civil affairs bureau, judicial authority etc.
Street bureau or Town (*zhen*)	"Leading group for the management of temporary residents"	Similar
Residents/Villagers committee	"Managing group for the floating population"	Mostly public security and housing management staff
Households		One "assisting hukou manager" per every 80 households in public housing (state or collectively owned) and per every 50 households in privately owned houses (mainly houses for leasing owned by local villagers)

dependent on the money collected from the temporary population. Therefore, extortion of fees and illegal practices are unavoidable (Guo, Li and Lin, 1999).

Furthermore, it is not only the village leaders who are involved in illegal rental activities; household assistant managers, who are often house owners, try to make unreported profits.

Even if the administrative system were not corrupt, because of the extremely high mobility of the floating population it is nearly impossible to exert proper control. For example, since the fees and taxes vary among the various towns and villages, floating people apply for certificates in places where they are the least expensive, but they live in different administrative units.

Conclusion

Given the present situation, there is clearly no simple solution for such a very complex and sensitive matter.

The basic requirement, however, seems to be that the legal civil rights of every migrant citizen should be observed (see also Zhou Daming, 1997b). Even if the Guangzhou municipal government recognizes the potential the transient population presents to generate income for the city at large, the performance of lower-level enforcement personnel is mostly insufficient (see also Solinger, 1999, p. 89). Instances of violation of law, extortion of money, harsh treatment, and illegal arrests still occur too often. If the personal safety of the migrant population were better protected, the general urban security situation would improve as well.

One of the major administrative obstacles seems to be the spatial coexistence of urban and rural residents' administrative bodies and so-called mass organizations under the jurisdiction of a city district. Therefore, the introduction of a unified management system, which would involve the abolition of the villagers' committees, would certainly improve the monitoring of the migrant population.

Coordination among different governmental agencies at both the same and different levels is insufficient. In general, the Public Security Bureau complains of the lack of support from other offices, bureaus and agencies, which often are unwilling to be involved in the administration of the floating population. Therefore, there are repeated demands by the PSB that other agencies should better cooperate and support its tasks. The very common top-down type of administrative approach seems to make close cooperation between different departments more difficult. The "front-line" governmental agencies, such as street bureaus, town governments or street level police stations (*pai chu suo*), always complain that the high-level bureaus load too many duties onto them without a corresponding authorisation.

Better management of the transient population is an urgent and important task, as is closer control and steering of the urban sprawl. So far suburban development has taken a chaotic course (Wu and Yeh, 1999). A better infrastructure (roads, sewage system) or the establishment of selected sub-centres will be required to coordinate the preservation of cultivated land and to protect traditional village landscapes.

The idea of resettling temporary migrants in far suburban enclosed camps, which seems to be a preferred solution to some planners and security officials, can certainly not be regarded as an attempt to integrate the transient population into urban society (Guo, Li and Lin, 1999).

The problem of the transient population can only be solved when the authorities and all members of the urban society regard these people as an indispensable and complementary part of the urban economy and life.

References

Chan, R. C. K. and Chaolin Gu. "Reforms of Metropolitan Development in Guangzhou Municipal City." in *Economic and Social Development in South China*, edited by S. MacPherson and J. Y.S. Cheng, pp. 281–305. Cheltenham: E. Elgar, 1996.

Chan, Sau-Hung, June. *Population Mobility and Government Policies in Post-Mao China*. M.Phil. Thesis , University of Hong Kong, 1996.

Chen, Xiao-ping. *Several Considerations Regarding the Concentrated Management of Chu Zu Wu*. Public Security Bureau, Research and Investigation Department. Guangzhou. Paper delivered to the "Symposium on the Management of *Chu Zu Wu* and *Wailai Renkou*," November 1999 (in Chinese).

Guangzhou Statistical Yearbook, various years.

Guo, Yanhua, Li, Ruihua, and Lin Wen-sheng. Research Office of the Public Security Bureau. *Several Opinions and Reflections on the Issue of Normalising the Management of Chu Zu Wu*. Paper delivered to the "Symposium on the Management of *Chu Zu Wu* and *Wailai Renkou*," November 1999 (in Chinese).

Lo, Fu-chen and P. J. Marcotullio. "Globalisation and Urban Transformations in the Asia-Pacific Region: A Review." *Urban Studies*, 37.1 (2000), pp. 77–111.

Public Security Bureau, Tianhe District. *Definition of the Concept of Chu Zu Wu*.

Paper delivered to the "Symposium on the Management of *Chu Zu Wu* and *Wailai Renkou*," November 1999 (in Chinese).

Research Department, Household Management Department, Public Security Bureau. *Status Quo of the Chu Zu Wu Management and Measures to Solve Related Problems.* Paper delivered to the "Symposium on the Management of *Chu Zu Wu* and *Wailai Renkou*," November 1999 (in Chinese).

Solinger, D. J. *Contesting Citizenship in Urban China: Peasant Migrants, the State, and the Logic of the Market.* Berkeley: University of California Press, 1999.

Study Group on Administrative Problems. *Administrative Problems in the Urban-rural Transitional Zone in Baiyun District, Guangzhou.* Faculty of Management Engineering, South Chinese Polytechnic of Construction, Guangzhou, 1999 (in Chinese).

Wu, Fulong. *Changes in the Urban Spatial Structure of a Chinese City in the Midst of Economic Reforms — A Case Study of Guangzhou.* Ph.D. Thesis, University of Hong Kong, 1995.

Wu, Fulong and A. Gar-on Yeh. "Urban Spatial Structure in a Transitional Economy: The Case of Guangzhou, China." *APA Journal,* 65.4 (1999), pp. 377–394.

Xu, Jiang. *Development Concepts and Land Use Planning Mechanisms in China: A Case Study of Guangzhou.* Ph.D. Thesis, University of Hong Kong, 1999.

Yang, C. K. *A Chinese Village in Early Communist Transition.* Cambridge: The MIT Press, 1959.

Yeh, A. Gar-on and Fulong Wu. "The New Land Development Process and Urban Development in Chinese Cities." *International Journal of Urban and Regional Research,* 20.2 (1996), pp. 330–353.

Zhou, Daming. "On Rural Urbanisation in China." In *Farewell to Peasant China,* edited by G. E. Guldin, pp. 13–46. Armonk: Sharpe, 1997a.

Zhou, Daming. "Investigative Analysis of 'Migrant Odd-Job Workers' in Guangzhou." In *Farewell to Peasant China,* edited by G. E. Guldin, pp. 227–247. Armonk: Sharpe, 1997b.

Zhou, Yi-Xing. *The Prospects of International Cities in China.* Paper presented at the International Conference on Urban Development in China: Last Half Century and into the next Millennium, 6–9 December 1999, Zhongshan/Guangdong.

6

Trends and Causes of Uneven Regional Development in the Zhujiang Delta

Jianfa Shen, Kwan-yiu Wong and David K. Y. Chu
Department of Geography, The Chinese University of Hong Kong

Introduction

The Zhujiang delta region is located in the central part of Guangdong province. In this paper, it refers to the Zhujiang Delta Economic Region formally designated by the government of Guangdong province in 1994 (Lu, 1995). In 1998 the region consisted of twenty-eight area units, including six urban areas of six prefecture-level cities that supervise other county-level units, three prefecture-level cities, sixteen county-level cities and three counties (Figure 1). The total area of the region is 41698 km^2, with a population of 22.38 million in 1998.

The Zhujiang delta region has experienced rapid growth since the early 1980s and has become one of the main economic growth centers in China. Its GDP accounted for 7.48 percent of mainland China's total GDP in 1998, at 583.3 billion RMB (Renminbi) (US$70.36 billion). This was about 43

This research is funded by the Research Grant Council of Hong Kong, RGC Reference No. CUHK4017/98H. An earlier version of this paper was presented at the International Workshop on Resource Management, Urbanization and Governance in Hong Kong and the Zhujiang Delta, jointly organized by United College and the Geography Department, The Chinese University of Hong Kong, 23–24 May 2000. Helpful comments from the workshop participants are gratefully acknowledged.

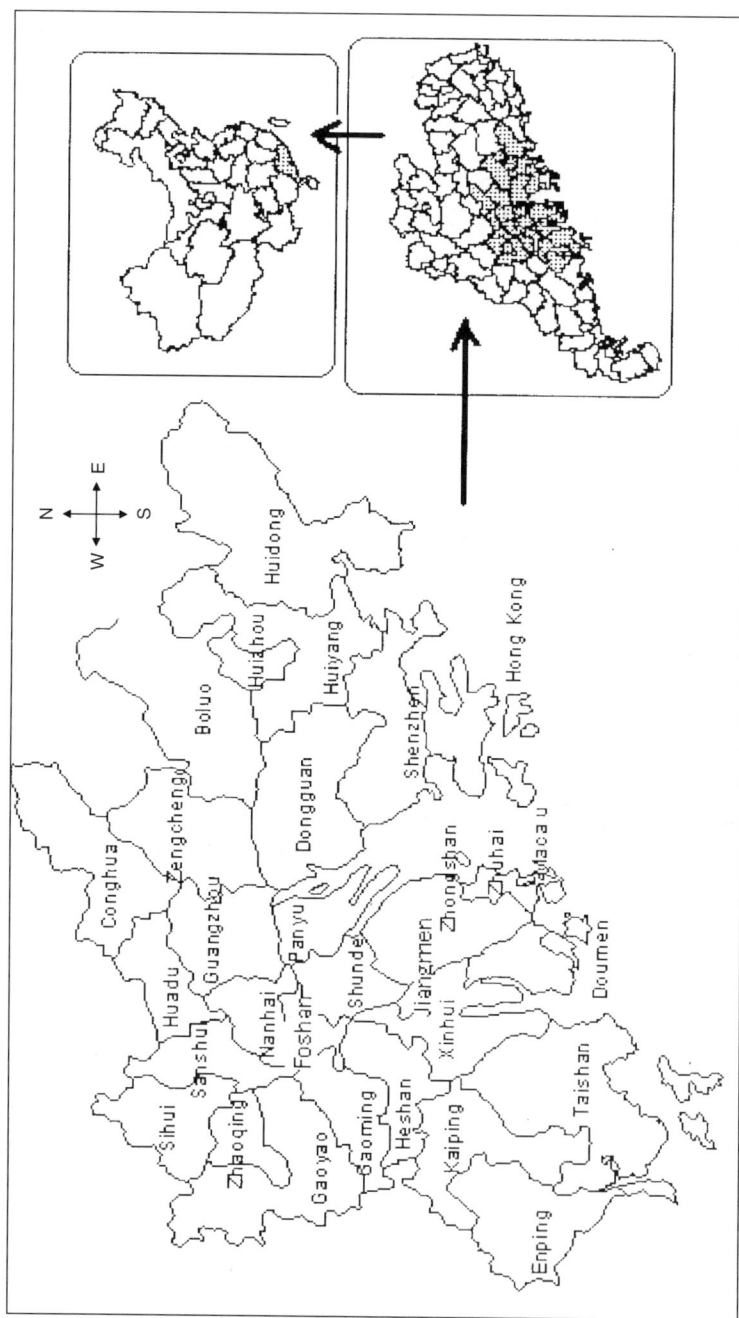

Figure 1. County-level area units in Zhujiang delta in 1998

percent of the GDP in Hong Kong in the same year. The Zhujiang delta had a high per capita GDP of 26,075 RMB (US$3145) in 1998, well over the national average, but still far behind the per capita GDP of US$ 24,449 in Hong Kong.

The implementation of the economic reforms and open-door policy, the coastal development strategy and non-local direct investment (NLDI, this term is more accurate than foreign direct investment, FDI, as much investment is from Hong Kong) have been the driving forces behind such rapid growth. Direct investment from Hong Kong and Macau has been particularly important in the Zhujiang delta. Various studies have been conducted on the Zhujiang delta. Some citations will be illustrative here. Xu and Li (1990) and Xu (1993) examine the changing patterns and distribution of urbanisation, demonstrating the urbanization process under China's open-door policy. Chan (1995) comprehensively analyzes major dimensions of recent development in the region. Sung *et al.* (1995) analyze the rapid development in the Zhujiang delta region, focusing on economic changes up to the early 1990s, while Yeung and Chu (1998) survey overall development in Guangdong province. Wang (1997) examines the important role of local government on development in the region, while Eng (1997) and Sit and Yang (1997) emphasize the important role of foreign investment in the recent urbanization in the region. Shen *et al.* (2000) examine the diffusion trends of FDI within the Zhujiang delta region. Shen (1999) and Shen *et al.* (1999) look at the rapid urban growth of Shenzhen and the driving forces behind it. Adjacent to Hong Kong across the Shenzhen River, the city has a population of more than 4 million. Both the innovative institutional reforms in the city and the external investment and labour migration are found to be important factors in the rise of Shenzhen city. Soulard (1997) carries out a preliminary analysis of urbanization in the Zhujiang delta using some county-level data. Lin (1997) comprehensively reviews the processes behind the rapid development in the Zhujiang delta but the empirical study focuses on rural industrialization, transport and the influence of Hong Kong. Henderson (1991), Smart and Smart (1991) and Sung (1998) examine the economic relations between Hong Kong and South China. Yeung (1999) and Li *et al.* (1996) examine issues of regional co-operation and integration between Hong Kong and the Zhujiang delta region, indicating the huge potentials and major challenges ahead. Chan (1998), Li (1998) and Chu *et al.* (2000) broadly assess regional development and urbanization in the region, indicating the need for regional integration and co-operation within the Zhujiang delta region. Overall, advantageous location, flexible reform and open policy "one step ahead" of other regions in China and large-scale non-local investment have been acknowledged to be important and the general pattern of urban and regional development in the region has been identified in previous studies. However, there is apparent lack of quantitative causal analysis of development in the region.

There has been growing interest in economics and geography on the long-term convergence in per capita income between sub-national regions and countries (Barro and Sala-i-Martin, 1995). There have been a number of studies on unbalanced development at the provincial level in China (Fan, 1992; 1995; Pannell, 1988; Zhou, 1993; 1996; Wei, 1996) or within a province such as Guangdong (Gu, *et. al.*, 2001). However, studies on the Zhujiang delta mostly focus on the overall growth of the region. Few studies have been conducted on the issue of unbalanced development within the region. Fan (1995) identifies the rising level of development in Shenzhen and Zhuhai in the early 1990s. Xu *et al.* (1999) finds decreasing regional disparities within the Zhujiang delta region in the 1990s. Few studies have attempted to explain the reasons behind balanced/unbalanced development in the Zhujiang delta. Li (1995) conducts an interesting study to analyze capital efficiency in the delta using relatively simple elasticity measures.

This paper will examine the trends and causes of uneven regional development in the Zhujiang delta region. The data on twenty-eight area units for this research have been collected from official sources (Guangdong Statistical Bureau, 1992; 1995; 1996; 1999). These data may not be as accurate as expected but are still broadly consistent with personal field observations. The analysis in the paper will cover the 1980–1998 period but detailed analysis of factor contributions will focus on 1980–1994, as the data for this period are relatively consistent. Some adjustments of the data have been made. First, GDP data were estimated on the basis of value-added data for the 1980–1994 period. Second, the data on capital stock for various areas were estimated from initial capital stock and investment data in various years. Third, the official population statistics usually only report the *hukou* population in various localities while there is a large non-*hukou* population in the Zhujiang delta region. This problem is avoided by using the employment data that report all employees in various economic sectors.

The rest of the paper is organized as follows. Section two will examine the trends of uneven regional development in the region during the 1980–1998 period. Section three will analyze the contributions of various factors to regional development using a production function approach. The conclusions will be presented in the final section.

Trends of regional development in the Zhujiang delta

The rapid economic growth in the Zhujiang delta has been well recognized. Table 1 presents per capita GDP in the Zhujiang delta, Guangdong and mainland China. In 1980, the level of development in Guangdong was very close to the national average, but the Zhujiang delta, with a sound agricultural sector, had a higher per capita GDP. However, Guangdong's level of development was well above the national average in 1998 and per capita GDP in the Zhujiang delta was more than two times that of Guangdong province. Guangdong and especially the Zhujiang delta have become one of the most developed regions in mainland China.

Table 1. Per capita GDP in the Zhujiang delta, Guangdong and mainland China 1980–1998 (at current prices)

Year	Zhujiang delta	Guangdong	Mainland China
1980 (RMB)	731	473	460
1985 (RMB)	1729	982	855
1990 (RMB)	4524	2395	1638
1995 (RMB)	18242	7973	4754
1998 (RMB)	26075	11154	6251
1998 (US$)	3145	1345	754

Sources: Shen *et al.* (1999); SSB (1999); Guangdong Statistical Bureau (1999).

Sub-regions experience different development paths under various internal and external circumstances. There has been relatively little analysis of the balanced/unbalanced development within the region. The main focus of the paper is to analyze the dynamic relations among various factors and regional economic growth among the sub-regions in the Zhujiang delta. Before conducting a causal analysis of regional economic growth in the next section, it will be of interest to examine the overall trend of uneven regional development in the region.

A major issue in urban and regional studies is whether the gap in the level of development between different areas is narrowing or widening (Friedmann and Alonso, 1975; Gilbert, 1976; Zhang and Shen, 1991; Alden and Boland, 1996). There has been growing interest in the long-term convergence in per capita incomes between regions (Barro and Sala-i-Martin, 1995). Two kinds of convergence have been identified. 'β-convergence' occurs if there is a negative correlation between the regional growth rate and the level of per capita income at the beginning of the period; 'σ convergence' occurs if the variance of relative per capita incomes decreases. This section will use a coefficient of variation similar to the 'σ convergence' approach and other complementary measures to examine trends in regional development in the region.

Initially, the analysis focuses on the per capita GDP during the 1980–1998 period based on 1990 prices. Table 2 presents the minimum, maximum, mean, standard deviation, and coefficient of variation (CV) of per capita GDP and the ratio of the maximum per capita GDP to the minimum per capita GDP. It is clear that all areas in the Zhujiang delta experienced rapid development in the period, as indicated by the rising level of the minimum per capita GDP. The minimum per capita GDP increased from 387 RMB in 1980 to 5,331 RMB in 1998. The maximum per capita GDP increased from 4,156 RMB in 1980 to 71,809 RMB in 1998. The average per capita GDP of twenty-eight areas increased from 1,036 RMB in 1980 to 14,670 RMB. The standard deviation, measuring the absolute regional difference, increased from 811 RMB to 12,708 RMB in the 1980–1998 period due to the rising level of per capita GDP.

The coefficient of variation might be a better measure of regional difference because it is not affected by the average level of per capita GDP. It shows that the regional differential initially increased in the 1980–1985

Table 2. Measures of unbalanced regional development 1980–1998
(Per capita GDP at 1990 prices)

Year	Minimum (RMB)	Maximum (RMB)	Mean (RMB)	Std. deviation (RMB)	CV	Max/Min
1980	387	4156	1036	811	0.78	10.74
1985	656	14691	2389	2718	1.14	22.38
1990	1581	29923	4859	5516	1.14	18.93
1991	1905	36523	6114	6814	1.11	19.18
1993	1685	53552	10034	10861	1.08	31.78
1994	2329	54706	11722	11525	0.98	23.49
1998	5331	71809	14670	12708	0.87	13.47

Sources: Processed from Guangdong Statistical Bureau (1992; 1995; 1999).
Notes: Std. deviation = standard deviation; CV = coefficient of variation.

period as the coefficient of variation increased from 0.78 in 1980 to a maximum of 1.14 in 1985. The regional differential then declined gradually from 1985 to 1998 and an 'σ convergence' occurred during this period. By 1998, the coefficient of variation was 0.87 smaller than that in 1985 but still greater than that in 1980. The ratio of the maximum per capita GDP to the minimum per capita GDP was the largest in 1993. This ratio in 1998 was smaller than that in 1993 but still greater than that in all previous years. It is clear that the regional differential increased in the initial period of rapid development in the region, but decreased or stabilized in the 1990s.

Figures 2–6 present per capita GDP for various areas in the region in 1980, 1985, 1990, 1994 and 1998 respectively. The detailed paths of economic growth in the various areas can be examined beyond simple summary convergence measures. Location quotients of GDP in terms of population are also calculated for various areas and are presented in Table 3 and Figure 7. If the location quotient of an area is greater than one, then the area has a greater share of GDP than its share of population and this means that its per capita GDP is above the average of the whole region.

There are similarities and changes in the spatial patterns of development in 1980 and 1998. According to Figure 2 and Table 3, Guangzhou, Zhuhai, Foshan and Jiangmen had the highest levels of per capita GDP in 1980 and they all had a location quotient of more than one. According to Figure 3, Shenzhen became an area with a high per capita GDP in 1985. Indeed, Shenzhen's per capita GDP was three times the region's average in 1985. In the 1985–1998 period, Shenzhen and Zhuhai had the highest per capita GDP in the region. The location quotients of GDP in Guangzhou and Jiangmen decreased, indicating that their per capita GDP became much closer to the average in the region. The relative level of development of other areas also increased or decreased during the 1980–1998 period. Using the location quotient of more or less than one in 1998 as a criteria to define more or less developed areas, four types of areas can be identified according to the trends of their location quotients in the Zhujiang delta region (Table 3).

Type A (more developed growing areas): areas of this type show an increase in the location quotient and the relative level of development and their per capita GDP was above the regional average by 1998. There are three sub-types: sub-type A_1 (most developed rapidly growing areas); sub-type A_2 (more developed rapidly growing areas) and sub-type A_3 (more developed slowly growing areas).

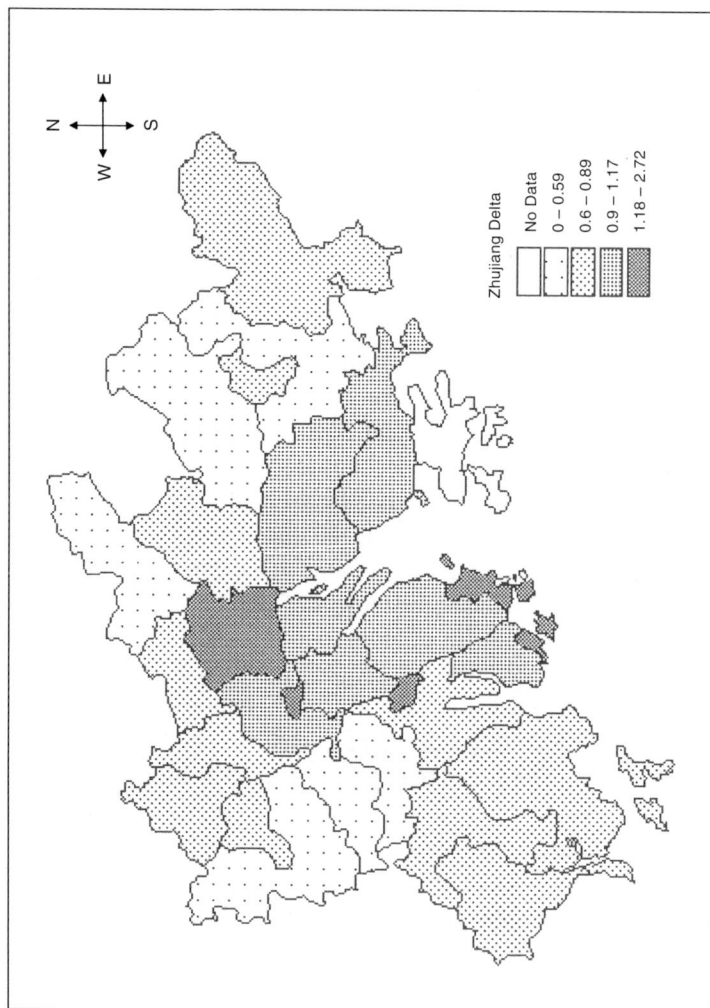

Figure 2. GDP per capita in Zhujiang delta in 1980 (thousand RMB)

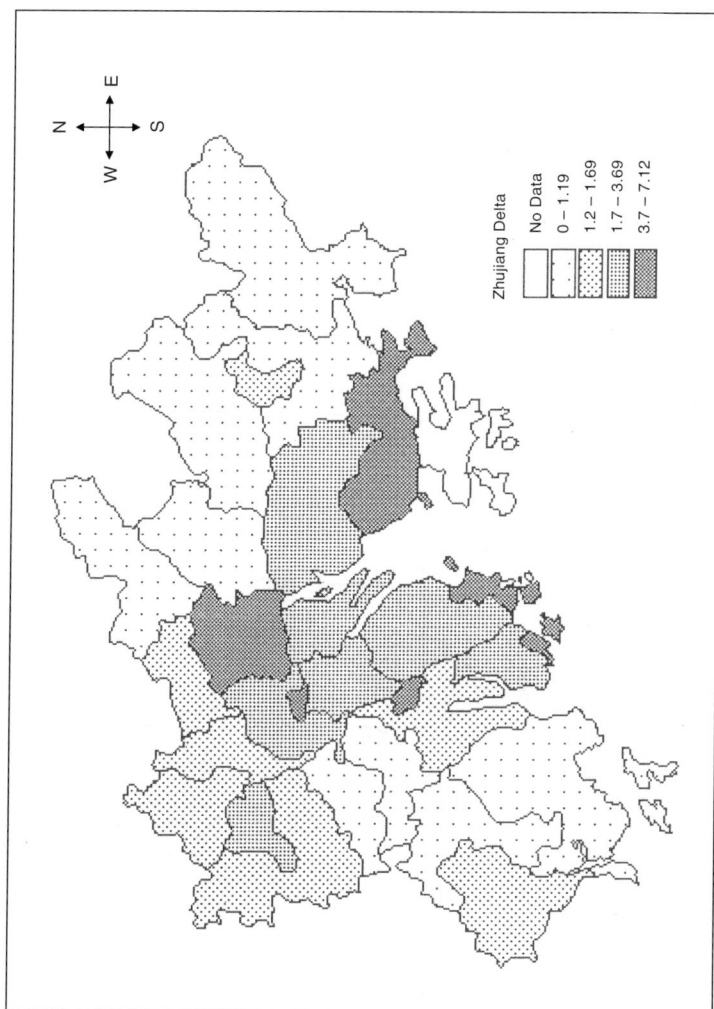

Zhujiang Delta

No Data
0 – 1.19
1.2 – 1.69
1.7 – 3.69
3.7 – 7.12

Figure 3. GDP per capita in Zhujiang delta in 1985 (thousand RMB)

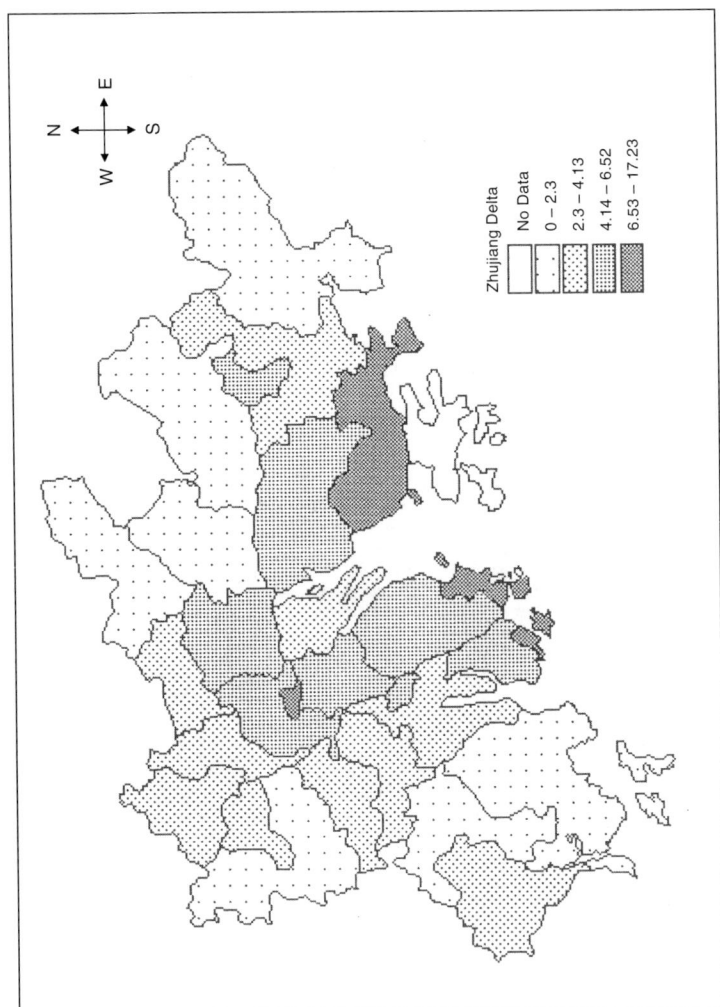

Figure 4. GDP per capita in Zhujiang delta in 1990 (thousand RMB)

Zhujiang Delta
- No Data
- 0 – 2.3
- 2.3 – 4.13
- 4.14 – 6.52
- 6.53 – 17.23

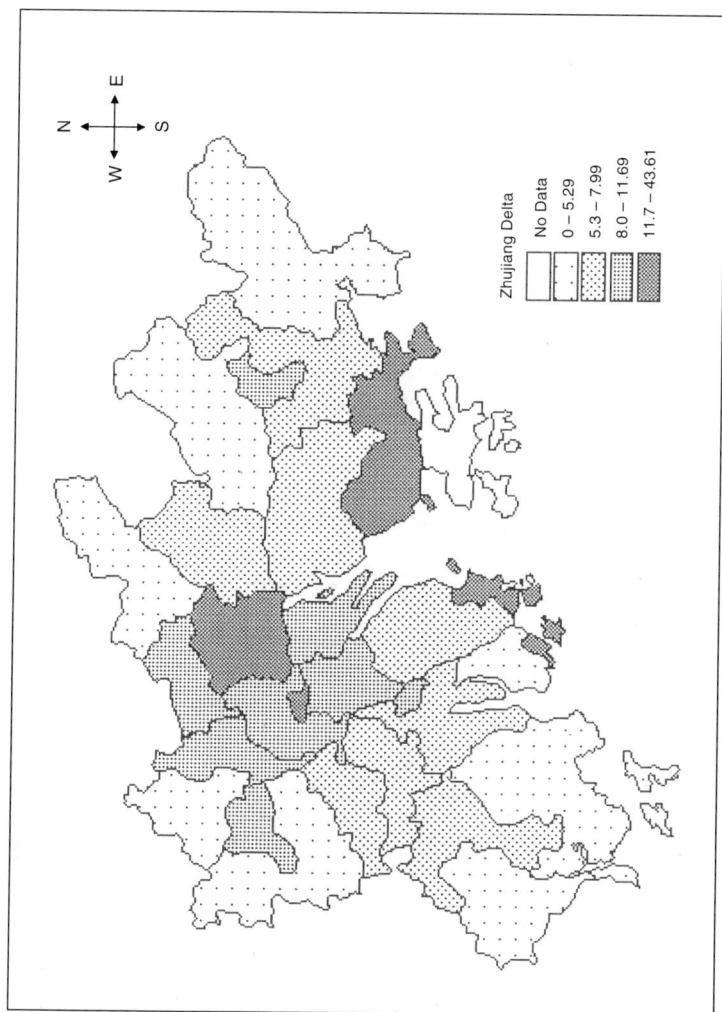

Figure 5. GDP per capita in Zhujiang delta in 1994 (thousand RMB)

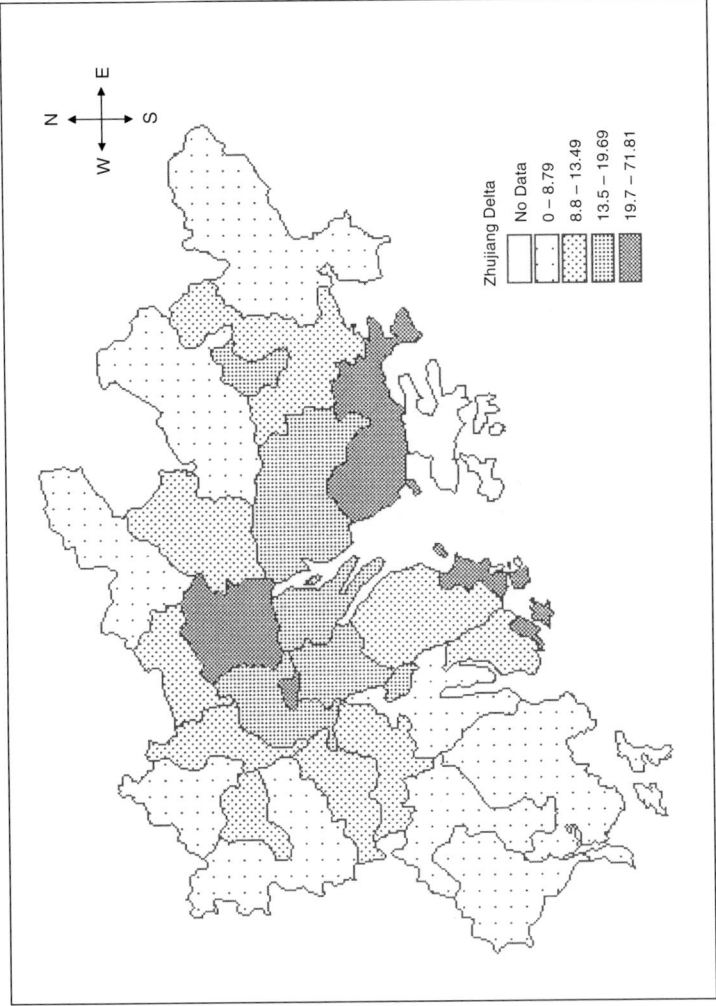

Figure 6. GDP per capita in Zhujiang delta in 1998 (thousand RMB)

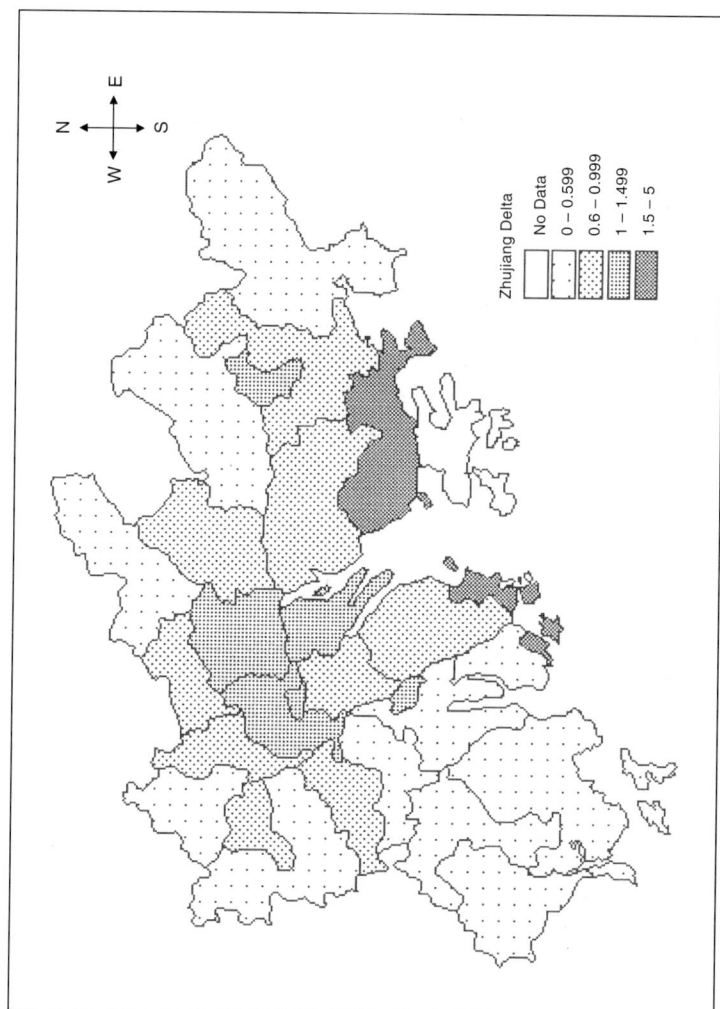

Figure 7. Location quotient of GDP in Zhujiang delta in 1998

Table 3. Location quotient of GDP in the Zhujiang delta 1980–1998

Area	1980	1998	Area type
Guangzhou	2.30	1.24	C
Huadu	0.58	0.81	B_1
Conghua	0.37	0.40	B_2
Zengcheng	0.59	0.61	B_2
Panyu	0.99	1.05	A_3
Shenzhen	0.92	4.31	A_1
Zhuhai	1.10	2.13	A_1
Doumen	0.99	0.53	D_2
Huizhou	0.53	1.10	A_2
Huidong	0.63	0.41	D_2
Huiyang	0.46	0.60	B_1
Boluo	0.33	0.35	B_2
Dongguan	0.79	0.91	B_1
Zhongshan	0.77	0.73	D_1
Jiangmen	1.71	1.11	C
Xinhui	0.57	0.41	D_2
Taishan	0.51	0.34	D_2
Kaiping	0.51	0.48	D_1
Enping	0.66	0.32	D_2
Heshan	0.49	0.58	B_1
Foshan	1.62	1.18	C
Nanhai	0.95	1.01	A_3
Shunde	0.97	0.94	D_1
Gaoming	0.37	0.62	B_1
Sanshui	0.51	0.64	B_2
Zhaoqing	0.73	0.72	D_1
Gaoyao	0.44	0.53	B_1
Sihui	0.54	0.44	D_2
Total	1.00	1.00	

Source: Calculated by the authors.

Notes: A: more developed growing areas

      A1: most developed rapidly growing areas

      A2: more developed rapidly growing areas

      A3: more developed slowly growing areas

   B: less developed growing areas

      B1: less developed rapidly growing areas

      B2: less developed slowly growing areas

   C: more developed relatively declining areas

   D: less developed relatively declining areas

      D1: less developed relatively slowly declining areas

      D2: less developed relatively rapidly declining areas

Type B (less developed growing areas): areas of this type show an increase in the location quotient and the relative level of development but their per capita GDP was still less than regional average by 1998. There are two sub-types: sub-type B_1 (less developed rapidly growing areas) and sub-type B_2 (less developed slowly growing areas).

Type C (more developed relatively declining areas): areas of this type show a decrease in the location quotient and the relative level of development but their per capita GDP was still above the regional average in 1998. Guangzhou, Jiangmen and Foshan belong to this type. They are more developed areas in the region but their relative level of development has declined.

Type D (less developed relatively declining areas): areas of this type show a decrease in the location quotient and the relative level of development and their per capita GDP was also below the regional average in 1998. There are two sub-types: sub-type D_1 (less developed relatively slowly declining areas) and sub-type D_2 (less developed relatively rapidly declining areas).

According to the location quotients in 1998 in Figure 7, Shenzhen and Zhuhai were the most developed areas in the region, followed by Guangzhou, Panyu, Huizhou, Jiangmen, Foshan and Nanhai which had a per capita GDP above the regional average. These areas are mostly located in the core of the Zhujiang delta and either experienced a rapid increase in their relative level of development, such as Shenzhen, Zhuhai and Huizhou, or had a high level of development in the early 1980s, such as Guangzhou, Panyu, Jiangmen, Foshan and Nanhai.

The changing spatial patterns of development result from the growth rates of per capita GDP in the various areas. Figures 8–11 present the average annual GDP growth rates for the periods 1980–1985, 1985–1990, 1990–1994 and 1994–1998 respectively. For the region as a whole, the annual growth rates were 16.65 percent, 15.97 percent, 30.24 percent and 14.71 percent for these periods respectively. The growth rate in the late 1980s was slightly slower than that in the early 1980s. The growth rate in the early 1990s was the most impressive because China had intensified its economic reform programme. The growth rate in the late 1990s was the slowest among the four periods due to the loss of a policy advantage and increasing competition from other regions of the country.

Some areas clearly experienced a high economic growth rate of over 30 percent a year during various periods. Shenzhen and Zhuhai, two of

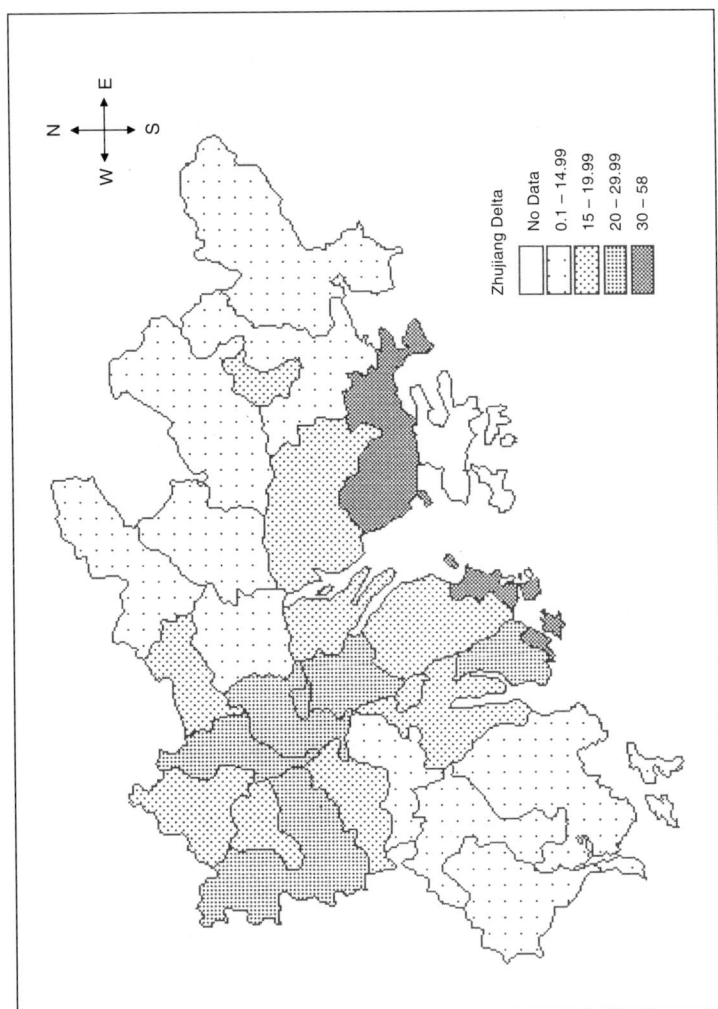

Figure 8. Annual GDP growth rate in Zhujiang delta in 1980–1985 (%)

Legend:

Zhujiang Delta

- No Data
- 0.1 – 14.99
- 15 – 19.99
- 20 – 29.99
- 30 – 58

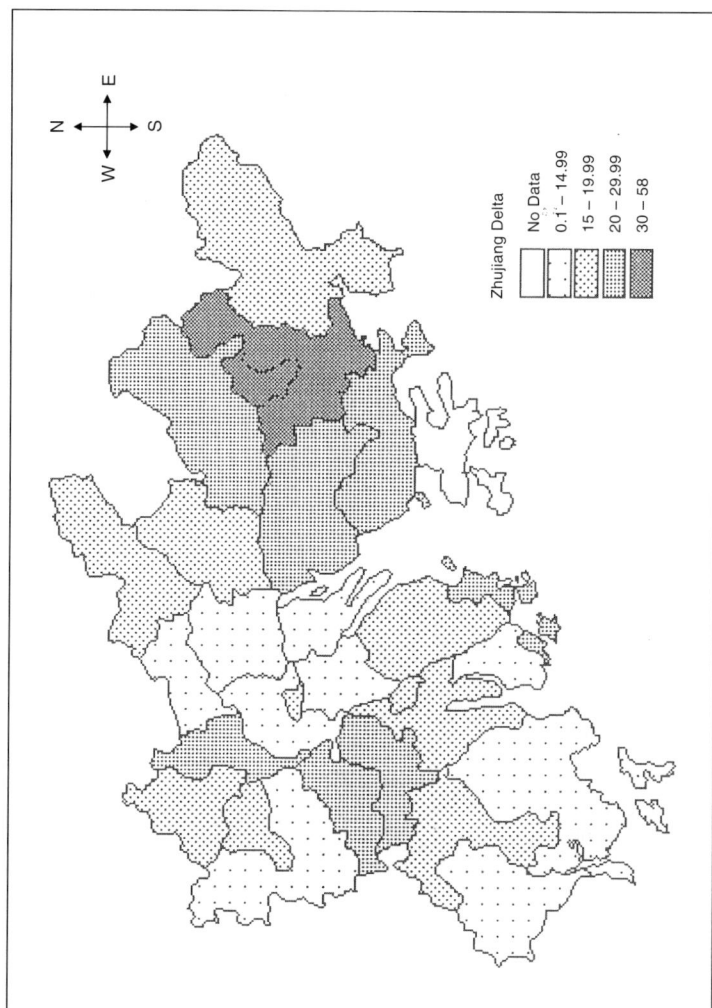

Figure 9. Annual GDP growth rate in Zhujiang delta 1985–1990 (%)

Zhujiang Delta

No Data
0.1 – 14.99
15 – 19.99
20 – 29.99
30 – 58

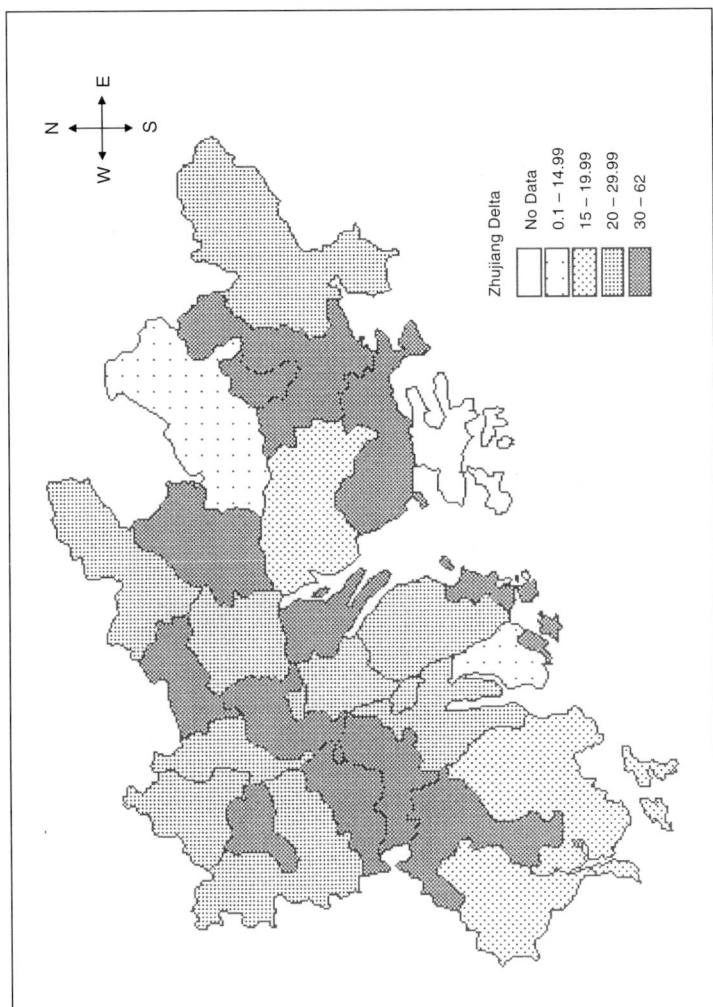

Figure 10. Annual GDP growth rate in Zhujiang delta 1990–1994 (%)

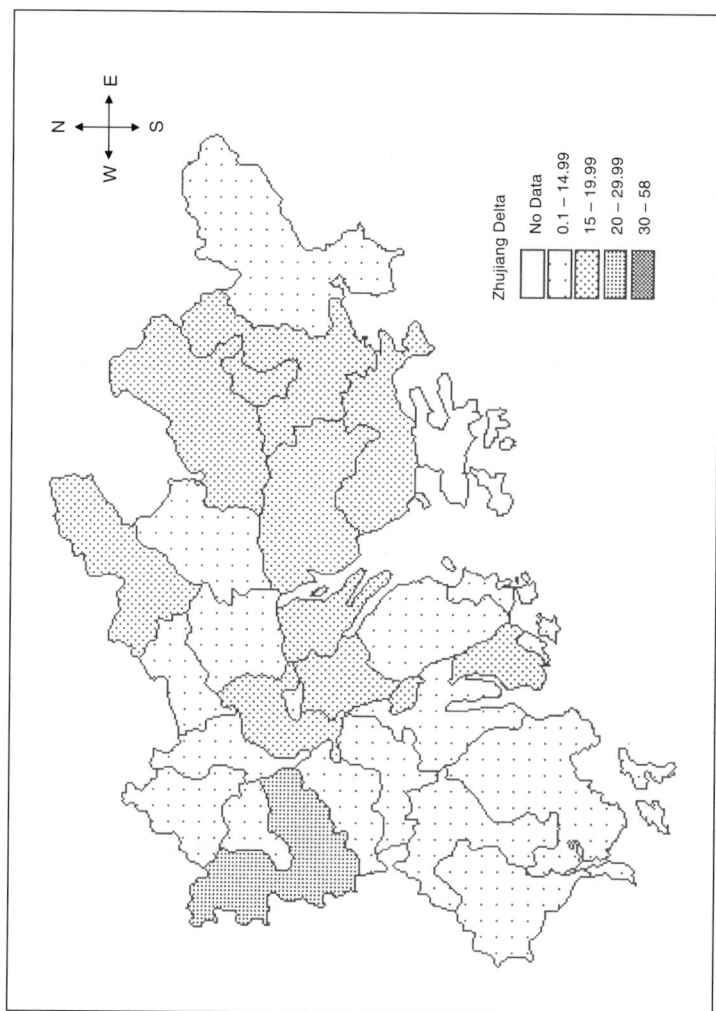

Figure 11. Annual GDP growth rate in Zhujiang delta 1994–1998 (%)

Zhujiang Delta

- No Data
- 0.1 – 14.99
- 15 – 19.99
- 20 – 29.99
- 30 – 58

China's four initial special economic zones, had annual GDP growth rates of over 30 percent in the first period (1980–1985) (Figure 8). Huizhou and Huiyang achieved growth rates of 30 percent per year in the second period (1985–1990) (Figure 9). As China's economic reforms and open-door policies were strengthened in the early 1990s, twelve areas in the Zhujiang delta achieved annual GDP growth rates of over 30 percent in the third period (1990–1994) (Figure 10). Only five areas, Doumen, Boluo, Dongguan, Taishan and Enping, had GDP growth rate of less than 20 percent per year in the 1990–1994 period. Most of these areas are located in the eastern and southeastern part of the region. Nevertheless, the smallest growth rate in Boluo was still 11.54 percent per year, indicating strong economic growth in the region during the 1990–1994 period. Even the slowest growing area in the Zhujiang delta had a growth rate well above the national average. During the 1994–1998 period, economic growth in various areas slowed down. Only Gaoyao achieved a growth rate of over 20 percent per year during this period. There were also eleven areas, Conghua, Panyu, Shenzhen, Doumen, Huizhou, Huiyang, Boluo, Dongguan, Jiangmen, Nanhai and Shunde, with GDP growth rates of 15 percent per year.

The evolving spatial economic structure in the Zhujiang delta region is also revealed through an analysis of the economic strength of the various areas in terms of their GDP shares in the total GDP of the region. Table 4 presents the GDP shares in the region during the 1980–1998 period.

Guangzhou was the dominant regional economic centre in the Zhujiang delta before the 1980s. In 1980, at the beginning of the reform period, Guangzhou accounted for 42.8 percent of the total GDP of the Zhujiang delta region, although only 18.86 percent of the region's population lived in the city. In 1980, there were only five other areas that accounted for 4–6 percent of the region's total GDP. These areas were Panyu, Dongguan,

Table 4. Share of GDP in the Zhujiang delta region 1980–1998

Area	1980	1985	1990	1994	1998
Guangzhou	42.80	34.23	24.44	23.69	22.20
Shenzhen	1.82	8.19	13.56	19.00	22.04
Top 8 cities	69.16	70.29	67.79	69.89	71.83
Other areas	30.84	29.71	32.21	30.11	28.17
Total	100	100	100	100	100

Source: Calculated by the authors.

Zhongshan, Nanhai and Shunde. These five strongest areas together only accounted for 23.68 percent of the region's GDP.

Regional development in the Zhujiang delta was characterized by the rise and fall of Shenzhen and Guangzhou respectively after 1980. The GDP share of Guangzhou declined rapidly from 42.80 percent in 1980 to only 22.20 percent by 1998. The GDP share of Shenzhen in the region increased dramatically from only 1.82 percent in 1980 to 22.04 percent by 1998. Shenzhen became the second largest economic power in the region in 1985, even though its economic power was still only one-fourth of that of Guangzhou. By 1998, Shenzhen's GDP share was very close to that of Guangzhou and was significantly greater than that of the other areas in the region.

A new spatial economic structure with two economic centres was finally formed in the region. The GDP shares of Guangzhou and Shenzhen in the region were as high as 44.24 percent while the third largest economic power, Dongguan, only accounted for 6.08 percent of the region's total GDP in 1998. Zhuhai's GDP share increased dramatically from only 0.86 percent in 1980 to 3.02 percent in 1990 and it has been stable since then. It has become an important economic power but not a major economic centre in the region. Relative economic strength also changed in other areas, but these changes were not as dramatic as those in Guangzhou, Shenzhen and Zhuhai. All five areas, Panyu, Dongguan, Zhongshan, Nanhai and Shunde, with a GDP share over 4 percent in 1980, retained this position in 1998.

The next section attempts to identify the contribution of capital, labour and other factors to the economic growth in various areas in the region.

Causes of regional development: A production function analysis

The relation between various factors and regional economic growth is an interesting problem. Figure 12 presents a simplified regional system. The regional economy consists of basic and non-basic sectors. The performance of the basic sector is crucial in determining a region's competitiveness in the national and global economy. Exports of goods and services by the basic sector allow a region to import important technology and resources as well as quality consumer goods and services. Non-local and foreign investment and external relations are mainly involved with the basic sector. Growth in the basic sector can also stimulate the non-basic sector through

Figure 12. A simplified regional system

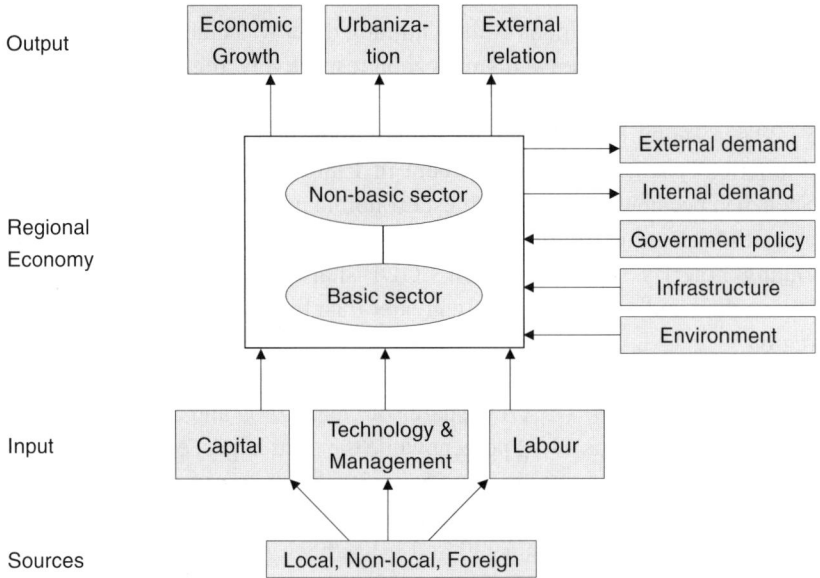

a multiplier effect. On the other hand, the non-basic sector exists to meet the internal needs of the region, especially property and consumer services that can not be imported. An increase in the productivity of the non-basic sector can also increase the GDP in the region and improve the living standards of the residents. When an urban or regional economy becomes more sophisticated, the share of the non-basic sector will become much greater than that of the basic sector. However, the non-basic sector is also very sensitive to the performance of the basic sector. When the basic sector declines, the non-basic sector will also be affected through a multiplier effect. This is why a government pays much attention to the basic sector, interregional and international imports and exports.

The basic inputs in the regional economy include capital, labour, technology and management that may be from local, non-local or foreign sources. Internal and external demands will affect the regional output. The regional economy also needs the support of the infrastructure and the environment. The performance of the regional economy will also be affected by government policies that regulate, constrain or facilitate regional economic activities. The result of regional economic performance may be presented directly in terms of indicators of economic growth, external trade and exports, or indirectly in terms of the progress of urbanization and the changing living standards of residents.

In the Zhujiang delta, the role of non-local investment in regional development has been emphasized, especially in terms of the scale of non-local investment in the region in comparison with other regions. Table 5 presents the non-local capital, or NLDI, in the Zhujiang delta, Guangdong and mainland China. In the early 1980s and 1990s, Guangdong received about 40 percent of the NLDI flow to mainland China. By 1998, other regions increased their share of NLDI but Guangdong still received 26.4 percent of the total NLDI flow to mainland China. During the 1979–1998

Table 5. Non-local capital (NLC) and non-local direct investment (NLDI) in Zhujiang delta, Guangdong and mainland China 1979–1998 (US$ billion)

Year	Guangdong NLDI	Mainland China NLDI	Guangdong share (%)	Delta NLC	Mainland China NLC	Delta share (%)
1979–1982	0.5	1.2	41.5	0.5	12.4	4.4
1983	0.2	0.6	40.9	0.3	2.0	13.5
1984	0.5	1.3	41.7	0.6	2.7	23.3
1985	0.5	1.7	30.3	0.8	4.6	18.1
1986	0.6	1.9	33.9	0.9	7.3	12.9
1987	0.6	2.3	25.8	0.8	8.5	9.8
1988	0.9	3.2	28.7	1.4	10.2	14.0
1989	1.2	3.4	34.0	1.5	10.0	14.8
1990	1.5	3.5	41.7	1.7	12.3	13.4
1991	1.8	4.4	41.4	1.9	11.6	16.8
1992	3.6	11.0	32.3	3.2	19.2	16.8
1993	7.5	27.5	27.3	6.4	39.0	16.4
1994	9.4	33.8	27.8	8.3	43.2	19.2
1995	10.2	37.5	27.1	8.6	48.1	17.8
1996	11.6	41.7	27.9	9.8	54.8	18.0
1997	11.7	45.2	25.9	11.8	64.4	18.3
1998	12.0	45.5	26.4	11.8	58.6	20.2
Total	74.4	265.7	28.0	70.5	408.9	17.2

Sources: SSB (1999) for Guangdong and China. For Zhujiang delta, Guangdong Statistical Bureau (1999) for 1995, 1997–1998; AEIPGD (1998) for 1996; Chan (1998) for 1979–1994.

period, Guangdong received a total NLDI of US$74.4 billion. About 14–20 percent of the total non-local capital flow to mainland China ended up in the Zhujiang delta. The region utilized a total of US$70.5 billion in the 1979–1998 period. The annual inflow of non-local capital to the region amounted to over US$10 billion each year in the late 1990s.

It is also interesting to examine the share of non-local capital in the total investment in Zhujiang delta (Table 6). It is clear that this share increased from only 5.5 percent in 1980 to 35 percent in 1998 for the region as a whole. Shenzhen and Zhuhai have had high shares of non-local capital since the early 1980s. Table 7 presents the share of non-local capital in the capital stock of the region. This share represents the accumulated results of non-local investment. In 1994, non-local capital accounted for 19.99 percent of the total capital stock in the region, increasing from 0.83 percent in 1980. In Shenzhen and Zhuhai, non-local capital accounted for 24.12 percent and 26.49 percent respectively of their total capital stock in 1994. It is likely that a certain amount of NLI (non-local investment) originated in mainland China but the investment was made via Hong Kong to take advantage of incentives for NLI. But it is not possible to identify the exact

Table 6. Share of non-local capital in total investment in
Zhujiang delta region 1980–1998 (%)

Area	1980	1985	1990	1991	1993	1994	1998
Guangzhou	3.04	5.97	8.69	12.97	13.61	23.69	25.28
Shenzhen	16.05	13.78	23.67	22.4	25.19	34.63	30.83
Zhuhai	24.88	14.53	23.09	28.66	27.29	34.43	49.95*
Zhujiang Delta	5.50	8.01	18.46	20.96	20.95	30.71	35.00

Sources: Shenzhen Planning Bureau (1999) and Guangdong Statistical Bureau (1999)
* Zhuhai Special Economic Zone in 1998 only.

Table 7. Share of non-local capital in fixed capital 1980–1994

Area	1980	1985	1990	1994
Guangzhou	0.53	3.69	5.88	12.32
Shenzhen	7.77	13.61	18.01	24.12
Zhuhai	13.42	14.93	17.73	26.49
Zhujiang delta	0.83	5.95	11.14	19.99

Source: Calculated by the authors.

amount of such investment as reliable data are unavailable. The NLI contribution may be slightly overestimated due to this problem.

Several studies have already analyzed and modelled regional economic growth, with an attempt to identify the contributions of capital, labour and productivity on regional growth (Beeson, 1987; Treyz, 1993; Spence and Vagionis, 1994). This paper adopts a well-known Cobb-Douglas production function to express the regional economy.

The following notations are used: Y is the regional GDP, L the labour force employed, K the capital stock, T the time, c the constant parameter, a the output elasticity of labour, d the output elasticity of capital, and v the rate of technical change.

Parameters c, a, d and v can be estimated using the empirical data. The production function can be expressed as:

$$Y = ce^{vT} L^a K^d \tag{1}$$

The following equation can be obtained by logarithmically differentiating equation (1) with respect to time:

$$\dot{Y} = a\dot{L} + d\dot{K} + v \tag{2}$$

Here $\dot{Y}$, $\dot{L}$ and $\dot{K}$ are the percentage growth rates of output, labour and capital. The above equation essentially decomposes the output growth rate into three components: labour growth (G_L), capital growth (G_K) and technical progress. It is noted that there is constant return to scale when $a + d = 1$, a decreasing return to scale when $a + d < 1$ and an increasing return to scale when $a + d > 1$. A constant return to scale should be used to calculate the contributions of labour and capital growth (Beeson, 1987), so that:

$$G_L + G_K = \frac{1}{a + d} (a\dot{L} + d\dot{K}) \tag{3}$$

$$G_L = \frac{1}{a + d} a\dot{L} \tag{4}$$

$$G_K = \frac{1}{a + d} d\dot{K} \tag{5}$$

A separate component of scale economy (G_S) may be calculated as:

$$G_S = (1 - \frac{1}{a + d}) (a\dot{L} + d\dot{K}) \tag{6}$$

The capital contribution in equation (5) can also be decomposed into contributions by local capital (G_{KL}) and non-local capital (G_{KNL}) using the share of non-local capital in the total capital stock formed in a period through investment (R_{KNL}), thus:

$$G_{KL} = (1 - R_{KNL}) \frac{1}{a + d} d\dot{K} \tag{7}$$

$$G_{KNL} = R_{KNL} \frac{1}{a + d} d\dot{K} \tag{8}$$

In this research, cross-sectional data for the 1980–1985 and 1985–1990 periods and the pooled data of the four years for the 1990–1994 period are used to estimate three sets of parameters of the production function respectively. A regional production function using all the data for the 1980–1994 period is also estimated. Table 8 presents the estimated results of the four regional production functions. The production equations were well estimated, with an adjusted R square of over 0.96. The F statistic was significant for all equations. The estimated labour elasticity and capital elasticity were also significant in all periods. The technical progress parameter was insignificant for the 1980–1985 period but it was significant for the other periods.

Table 8. Parameter estimates of regional production functions

Period	1980–85	1985–90	1990–94	1980–94
Number of samples	56	56	112	168
R	0.972	0.975	0.961	0.978
Adjusted R^2	0.942	0.949	0.922	0.955
F	297	339	437	1194
In constant	−3.126	−3.042	−1.744	−2.086
	(5.336)	(5.258)	(3.343)	(5.382)
v	−0.023*	0.073	0.088	0.064
	(1.500)	(5.770)	(4.944)	(10.320)
a	0.578	0.550	0.391	0.464
	(8.700)	(8.999)	(6.867)	(10.244)
d	.598	0.595	0.640	0.619
	(12.130)	(14.687)	(15.966)	(19.224)

Source: Calculated by the authors.

Notes: R = Correlation coefficient of regression equation; F = F statistic; v = parameter of technical progress; a = labour elasticity; d = capital elasticity;

* = insignificant at 0.05 level; t-values in brackets.

Table 9 presents selected results of the estimated contributions of various factors on economic growth in the Zhujiang delta as a whole and in Guangzhou, Shenzhen and Zhuhai. In the first period (1980–1985), 79.22 percent of regional economic growth in the Zhujiang delta was due to local investment, 16.49 percent due to the increased scale of the regional economy, 7.02 percent due to the increased labour force, and 7.47 percent due to NLI. There was no overall technical progress and its contribution of 10.2 percent was negative. This may mean that less advanced technology was employed in the region in the early 1980s, particularly in the collectively and privately owned sectors compared to that employed in the previous regional economy dominated by the state-owned sector. In Guangzhou the contribution of local capital was as high as 85.85 percent. In Shenzhen and Zhuhai the contribution of local capital was less than the average in the delta but it was still over 69 percent. Non-local capital contributed over 11 percent to the economic growth in these two cities.

In the second period (1985–1990), technical progress became an

Table 9. Percentage contributions to regional economic growth in Guangzhou, Shenzhen, Zhuhai and the Zhujiang delta by various factors 1980–1994 (%)

Area	Labour	Local capital	Non-local capital	Capital total	Scale economy	Technology
1980–1985						
Guangzhou	9.02	85.85	5.99	91.84	17.75	−18.61
Shenzhen	5.05	69.57	11.26	80.83	15.12	−1.00
Zhuhai	2.48	71.12	12.59	83.71	15.17	−1.36
Delta	7.02	79.22	7.47	86.69	16.49	−10.20
1985–1990						
Guangzhou	3.69	40.36	4.01	44.37	6.97	44.97
Shenzhen	30.53	30.89	8.32	39.22	10.11	20.14
Zhuhai	31.52	28.17	7.21	35.38	9.70	23.41
Delta	8.37	39.19	7.94	47.13	8.05	36.45
1990–1994						
Guangzhou	4.80	44.91	11.14	56.04	1.89	37.27
Shenzhen	39.76	26.76	11.31	38.07	2.41	19.75
Zhuhai	23.90	36.68	17.35	54.03	2.42	19.66
Delta	8.52	45.31	16.65	61.96	2.18	27.34

Source: Calculated by the authors.

important force underlying the economic growth, accounting for 36.45 percent of regional economic growth in the Zhujiang delta; 39.19 percent of the economic growth in the region was due to local investment; 8.05 percent was due to the increased scale of the regional economy; 8.37 percent was due to the increased labour force; and 7.94 percent was due to NLI. It is noted that the labour contribution was as high as 30 percent in both Shenzhen and Zhuhai, indicating the importance of migration and population growth to these growing cities.

In the third period (1990–1994), technical progress was slightly less important to economic growth than during the previous period, accounting for 27.34 percent of the regional economic growth in the Zhujiang delta. Similarly, the scale economy only accounted for 2.18 percent of the economic growth in the region. Local capital and NLI were more important than during the previous period. 45.31 percent of the economic growth in the region was due to the local investment and 16.65 percent was due to NLI. The contribution of the labour force was 8.52 percent, similar to during the previous period. The labour contribution was still very important in Shenzhen and Zhuhai (39.76 percent and 23.90 percent respectively). It is noted that the contribution of non-local capital in Shenzhen and Zhuhai was smaller than the average in the delta, indicating the diffusion of non-local capital in areas outside of the special economic zones. Technical progress was a significant factor in the economic growth of Guangzhou during this and the previous period. It accounted for 44.97 percent and 37.27 percent of the economic growth in 1985–1990 and 1990–1994 respectively.

The estimated contributions of various factors can be used to trace the reasons for increases or decreases in an area's economy. For example, it is found that the contributions of non-local capital investment and labour growth were relatively small in Guangzhou during the 1985–1994 period and this may be the main reason for the relative decline of Guangzhou's leading position in the delta region. On the other hand, rapid economic growth in Shenzhen and Zhuhai was related to a larger contribution by non-local capital during the 1980–1985 period and significant migration and labour growth during the 1985–1994 period.

Conclusion

This paper assesses the dynamics of spatial development in the Zhujiang delta region during the 1980–1998 period. The implementation of economic reforms and the open-door policy, the coastal development strategy and non-local direct investment have been the driving forces behind such rapid growth since the late 1970s. This paper examines the trends and causes of the uneven regional development in the Zhujiang delta region. The whole region shows that a trend of an inverted U-curve of regional differentials and 'σ convergence' occurred in the early 1990s. Clear spatial transformation occurred in the region and the positions of Shenzhen and Zhuhai were strengthened. The initial single center spatial structure dominated by Guangzhou was replaced by a structure with the two leading centers of Guangzhou and Shenzhen during the 1980–1998 period. Although there has been stable growth in Guangzhou, its relative strength has been significantly weakened by the rapid rise of Shenzhen.

The differential economic growth among various cities/counties is the result of the different inputs of labour and capital and technical progress. Using a decomposition approach based on regional production functions, the relative contributions of various factors, such as local capital, non-local investment, labour force growth and technical progress, to the economic growth in the various areas in the region have been identified. It is clear that local investment has been the most important contributor to economic growth, although the region has attracted a large amount of NLI. Even in Shenzhen and Zhuhai, the two special economic zones designated in 1980 that are on the frontier of the Chinese open-door policy, NLI only contributed less than 18 percent of economic growth during the 1980–1994 period. It is clear that the process of local capital accumulation is also important to the rapid economic growth in the Zhujiang delta. Indeed, the share of savings and investment often accounts for a large portion of GDP in the region, resulting in the high economic growth rate.

The estimated contributions of various factors can be used to trace the reasons for increases or decreases in an area's economy. For example, it was found that the contributions of non-local capital investment and labour growth were relatively small in Guangzhou during the 1985–1994 period and this may be the main reason for the relative decline of Guangzhou's leading position in the delta region. On the other hand, rapid economic growth in Shenzhen and Zhuhai was related to a larger contribution of non-local capital during the 1980–1985 period and significant migration and labour growth during the 1985–1994 period.

The findings in this paper point to the importance of local capital accumulation in regional economic growth. More generally, location initiatives and local responses are also important to make best use of the unique opportunities offered by national policy and external capital. Without a quality labour force and enthusiastic local entrepreneurs and officials, the economic miracle of the Zhujiang delta would not have emerged. The same is true for non-local capital. Without the Hong Kong factor and NLI, economic growth in the region would not have been so rapid over the past two decades.

References

AEIPGD (Association of Economy and Investment Promotion, Guangdong). *Economic Investment Environment in Guangdong*. Guangzhou: Guangdong Education Press, 1998.

Alden, J. and P. Boland (eds.). *Regional Development Strategies: A European Perspective*. London: Jessica Kingsley Publishers and Regional Science Association, 1996.

Barro, R. J. and X. Sala-i-Martin. *Economic Growth*. New York: McGraw-Hill, Inc., 1995.

Beeson, P. "Total Factor Productivity Growth and Agglomeration Economies in Manufacturing, 1959–73." *Journal of Regional Science*, 27 (1987), pp. 183–199.

Chan, R. C. K. "The Pearl River Delta Region." In *Development in Southern China: A Report on the Pearl River Delta Region*, edited by J. Cheng and S. Macpherson, pp. 1–21. Hong Kong: Longman, 1995.

Chan, R. C. K. "Regional Development in the Yangtze and the Zhujiang Delta Regions." In *The Guangdong Development Model and its Challenges*, edited by J. Y. S. Cheng , pp. 43–79. Hong Kong: City University of Hong Kong Press, 1998.

Chu, D. K. Y., J. Shen and K. Y. Wong. "Regional/Urban Governance: A Prerequisite for Regional Development." In *New Prospect of Regional Development*, edited by X. J. Li and H. Qian, pp. 13–19. Kaifeng: Henan University Press, 2000.

Eng, I. "The Rise of Manufacturing Towns: Externally Driven Industrialization and Urban Development in the Pearl River Delta of China." *International Journal of Urban and Regional Research,* 21 (1997), pp. 554–568.

Fan, C. "Foreign Trade and Regional Development in China." *Geographical Analysis,* 24.3 (1992), pp. 240–256.

Fan, C. "Of Belts and Ladders: State Policy and Uneven Regional Development in Post-Mao China." *Annals of the Association of American Geographers,* 85.3 (1995), pp. 421–449.

Friedmann, J. and W. Alonso. *Regional Policy: Readings in Theory and Applications*. Cambridge: MIT Press, 1975.

Gilbert, A. (ed.). *Development Planning and Spatial Structure*. London: Wiley, 1976.

Gu, C., J. Shen, K. Y. Wong and F. Zhen. "Regional Polarization under the Socialist-Market System since 1978: A Case Study of Guangdong Province in South China." *Environment and Planning A*, 33 (2001), pp. 97–119.

Guangdong Statistical Bureau. *Zhujiang Delta Economic Statistics 1980–1991*. 1992.

Guangdong Statistical Bureau. *Zhujiang Delta Economic Statistics 1980–1994*. 1995.

Guangdong Statistical Bureau. *Statistical Yearbook of Guangdong 1996*. Beijing: China Statistical Publishing House, 1996.

Guangdong Statistical Bureau. *Statistical Yearbook of Guangdong 1999*. Beijing: China Statistical Publishing House, 1999.

Henderson, J. "Urbanization in the Hong Kong-South China Region: An Introduction to Dynamics and Dilemmas." *International Journal of Urban and Regional Research*, 15 (1991), pp. 169–179.

Li, H., L. Xu and W. Zhou (eds.). *The Regional Economy of Guangdong, Hong Kong and Macau at the Turn of the Century*. Guangzhou: Guangdong Higher Education Press, 1996.

Li, Kui Wai. "Capital Efficiency in the Pearl River Delta." In *Development in Southern China: A Report on the Pearl River Delta Region*, edited by J. Cheng and S. Macpherson, pp. 171–185. Hong Kong: Longman, 1995.

Li, Si-ming. "Pearl Riversville: A Survey of Urbanization in the Pearl River Delta." In *The Guangdong Development Model and its Challenges*, edited by J. Y. S. Cheng, pp. 81–109. Hong Kong: City University of Hong Kong Press, 1998.

Lin, G. C. S. *Red Capitalism in South China: Growth and Development of the Pearl River Delta*. Vancouver: UBC Press, 1997.

Lu, T. "The Pearl River Delta: Economic Wonder of the 80's and the Reinforced Vantage of the 90's." In *New Perspectives of the Economic Development of the Pearl River Delta*, edited by Zhujiang Delta Economic Development and Management Research Center, Zhongshan University, pp. 3–19. Guangzhou: Zhongshan University Press, 1995 (in Chinese).

Pannell, C. W. "Regional Shifts in China's Industrial Output." *Professional Geographer*, 40.1 (1988), pp. 19–32.

Shen, J. "Urbanization in Southern China: The Rise of Shenzhen City." In *Problems of Megacities: Social Inequalities, Environmental Risks and Urban Governance*, edited by A. G. Aguilar and I. Escamilla, pp. 635–648. Mexico City: Universidad Nacional Autonoma de Mexico, 1999.

Shen, J., D. K. Y. Chu and K. Y. Wong. "The Shenzhen Model: Forces of Development and Future Direction of a Mainland City near Hong Kong." In *Studies on the Regional Integration under the Model of "One Country Two Systems,"* edited by S. Ye, Y. Niu and C. Gu, pp. 112–131. Beijing: Sciences Press, 1999.

Shen, J., K. Y.Wong, K. Y. Chu and Z. Feng. "The Spatial Dynamics of Foreign Investment in the Pearl River Delta, South China." *The Geographical Journal,* 166.4 (2000), pp. 312–322.

Shenzhen Planning Bureau. *Economic and Social Development of Shenzhen — Reviews and Prospects 1998–1999.* Shenzhen: Haitian Press, 1999.

Sit, V. F. S. and C. Yang. "Foreign-investment-induced Exo-urbanization in the Pearl River Delta, China." *Urban Studies,* 34 (1997), pp. 647–677.

Smart, J. and A. Smart. "Personal Relations and Divergent Economics: A Case Study of Hong Kong Investment in South China." *International Journal of Urban and Regional Research,* 15 (1991), pp. 216–233.

Soulard, F. *The Restructuring of Hong Kong Industries and the Urbanization of Zhujiang Delta 1979–1989.* Hong Kong: The Chinese University Press, 1997.

Spence, N. A. and N. Vagionis. "Total Factor Regional Productivity in Greece." *Environment and Planning C: Government and Policy,* 12 (1994), pp. 383–407.

SSB. *Comprehensive Statistical Data and Materials on 50 Years of New China.* Beijing: China Statistical Publishing House, 1999.

Sung, Y. W. *Hong Kong and South China: The Economic Synergy.* Hong Kong: City University of Hong Kong Press, 1998.

Sung, Y. W., P. W Liu, R. Y. C. Wong and P. K. Lau. *The Fifth Dragon: The Emergence of the Pearl River Delta.* Singapore: Addison Wesley Publishing Co., 1995.

Treyz, G. I. *Regional Economic Modelling: A Systematic Approach to Economic Forecasting and Policy Analysis.* Boston: Kluwer Academic Publishers, 1993.

Wang, L. (ed.). *Economic Development and Local Government: A Study of the Zhujiang Delta Region.* Guangzhou: Zhongshan University Press, 1997.

Wei, Y. "Fiscal Systems and Uneven Regional Development in China, 1978–1991." *Geoforum,* 27.3 (1996), pp. 329–344.

Xu, X. Q. "Retrospect and Prospectus of Urbanization in the Zhujiang Delta." In *Urban and Regional Development in China: Looking into the 21st Century,* edited by Y. M. Yeung, pp. 369–384. Hong Kong: Hong Kong Institute of Asia-Pacific Studies, The Chinese University of Hong Kong, 1993 (in Chinese).

Xu, X. Q. and S. M. Li. "China's Open Door Policy and Urbanization in the Pearl River Delta Region." *International Journal of Urban and Regional Research,* 14 (1990), pp. 49–69.

Xu, X. Q., X. Yan, Y. Xu and J. Tian. "An Analysis of Regional Disparities in Guangdong Province in 1990s." In *Urban, Rural and Regional Development,* edited by Y. M. Yeung, D. Lu and J. Shen, pp. 453–474. Hong Kong: Hong Kong Institute of Asia-Pacific Studies, 1999.

Yeung, Y. M. "The Emergence of Pearl River Delta Mega Urban-region in a Globalizing Environment." *Occasional Paper* No. 90, pp. 1–24. Hong Kong: Hong Kong Institute of Asia-Pacific Studies, 1999.

Yeung, Y.M. and D.K.Y Chu (eds.). *Guangdong: Survey of a Province Undergoing Rapid Change.* Hong Kong: The Chinese University Press, 1998.

Zhang, C. and J. Shen. *Theory of Regional Science.* China Soft Sciences Series No.8. Huhai: HUST Press, 1991 (in Chinese).

Zhou, Q. "Capital Construction Investment and its Regional Distribution in China." *International Journal of Urban and Regional Research,* 17 (1993), pp. 159–177.

Zhou, Q. "Interprovincial Resource Transfers in China, 1952–90." *International Journal of Urban and Regional Research,* 20 (1996), pp. 571–586.

PART III

Resource Management

7

Sustainable Forestry in Hong Kong: Catalytic Effects of Acacia Plantations on the Invasion of Native Species

Kwai-cheong Chau and Pui-sze Au
Department of Geography, The Chinese University of Hong Kong

Introduction

Hong Kong used to support a climax vegetation of semi-deciduous monsoon forest (Thrower, 1975), which disappeared as a result of cutting before the mid-1850s (Dudgeon and Corlett, 1994). Forest rehabilitation was initiated thereafter, but all the plantations were destroyed during the Japanese occupation between 1942 and 1945. A massive reforestation programme was carried out after the war to protect the soil and to conserve water resources. These plantations are dominated mostly by the exotic species of *Acacia confusa, Acacia auriculiformis, Eucalyptus spp., Lophostemon confertus, Melaleuca quinquenervia* and *Pinus elliottii*. Local experience on the management of the old plantations is lacking, except for clearance of the understorey layer and its replacement with native tree species since 1998.

The plantations in Hong Kong can be divided into two broad categories,

The study was supported by the United College Campus Work Scheme and the Department of Geography of The Chinese University of Hong Kong. Assistance provided by the Agriculture and Fisheries Department is gratefully acknowledged. We are also grateful to Kelvin Lee, Maxwell Luk, K.P. Lui, K.H. Yeung and T.H. Cho for their assistance in the field work.

namely reforestation of severely disturbed areas such as the borrow areas and badlands, and enhancement planting on slopes that have suffered from fire and cutting. The objectives of rehabilitation planting are vaguely defined in the local context as a process to protect the soil and conserve water resources. Very often revegetation of the degraded areas is considered an end itself (Tsang, 1997), not a means to systematically restore the ecosystem as advocated by Lugo (1988). As borrow areas and badlands are impoverished in soil fertility, options are limited regarding the selection of species mix (Aber, 1993; Bradshaw, 1993). With a capacity to fix the atmospheric nitrogen and to augment the soil nitrogen supply, legumes are preferred to non-legumes in rehabilitation planting (Giller and Wilson, 1991; Kendle and Bradshaw, 1992). In Hong Kong, the exotic nitrogen-fixing legumes of *A. confusa* and *A. auriculiformis* have been extensively used to revegetate degraded slopes since the early 1960s and the mid-1980s, respectively. Zhuang and Yau (1998) investigated the floristic diversity of three types of plantations in Hong Kong, yet plantations established on severely disturbed environment were excluded. It is anticipated that increasingly more severely disturbed lands, such as the borrow areas, abandoned quarries, and landfills, will require restoration as a result of infrastructure and urban development. However, the development of the understorey layer in the restored borrow areas is least understood, not to mention management prescriptions needed for these plantations (Lugo, 1997).

Hill fires remain the greatest threat to local vegetation. Annually there are about 300 outbreaks of fire, consuming on an average 130,000 trees (Agriculture and Fisheries Department, 1980–1995). Where fires occur annually or twice every three years, the succession of vegetation is arrested at the grassland stage, which is relatively poor in species composition (Daley, 1975; Thrower and Thrower, 1986). Since the mid-1960s, enhancement planting of exotic species has been carried out on slopes that have suffered from such mild disturbances as fire and cutting. The purpose of enhancement planting is to accelerate recovery of the native forest through a nurse effect of the exotic species (Parrotta, 1993; Lamb, 1998). As propagules are abundant on these disturbed slopes, their protection against fires for a period of approximately 20 years will foster the natural succession of the native forest (Hodgkiss *et al.*, 1981). Therefore, enhancement planting of the grassy slopes with exotic species may become redundant and cost-ineffective.

More nitrogen-fixing legumes are increasingly used in reforestation and enhancement planting in Hong Kong. These legumes have an average lifespan of around 60 years, depending on the site quality and the intensity of management. The growth characteristics of these legumes with age, particularly *A. confusa*, and their nurse effect on native forest succession are least understood in the local environment. A survey of the floristic characteristics of these plantations not only provides necessary information to fill this knowledge gap but also a framework to review the goals of reforestation and enhancement planting. Therefore, this study sets out to investigate the floristic diversity of six plantations established between the mid-1960s and early 1990s, with the following three specific objectives: (a) to examine the growth characteristics of the overstorey vegetation, (b) to investigate the floristic composition of the understorey layer of the plantations, and (c) to review the objectives of reforestation and enhancement planting.

Study sites

Six sites were chosen for the present study, each dominated by either the monoculture of *Acacia confusa* and *Acacia auriculiformis* or the polyculture of the same species plus *Lophostemon confertus*. These plantations vary in age from 5 to 35 years, and were pit planted at 1.3–1.5 m apart by the Agriculture and Fisheries Department. After planting, the sites were maintained for 2 years, including weeding, beating up planting, fertilizing (twice per year) and fire protection (Agriculture and Fisheries Department, 1989). When the plantations were 4- to 5-year old, pruning and thinning were carried out to ensure the development of quality stands. The plantations were then left to develop on their own, without further treatment. The establishment of understorey species therefore represents *in situ* germination of the buried seeds and regeneration of the available propagules, as well as invasion from outside the sites. Of course, the regeneration mechanisms vary with the species and the severity of the disturbance (Oliver and Larson, 1996). As site conditions prior to the establishment of the plantations are largely conjectural, it is difficult to differentiate the original and invaded species of the understorey layer.

The first site (P5) is located in Tai Tong, northwest of the New Territories of Hong Kong. It is underlain by red yellow podzol derived from medium-grained granite and it was previously a gullied badland dominated by the

Table 1. Characteristics of the study sites

Site	Location (grid ref.)	Overstorey	Age* (years)	Spacing of planting*	Geology	Soil type	Elevation (m)	Av. slope gradient	Aspect	Earlier land use
P5	Tai Tong (952812)	A. auriculiformis A. confusa L. confertus	5	1.5 m	Granite	Red-yellow podzol	60–85	25°	Southwesterly	Borrow area
P6	Tai Lam (966783)	A. auriculiformis	6	1.5 m	Granite	Red-yellow podzol	270–300	23°	Northeasterly	Grassland with few shrubs
P13	Wan Tsai (263869)	A. confusa	13	NA	Granite	Red-yellow podzol	30–60	15°	Westerly	Borrow area
P14	Lau Shui Heung (089905)	A. auriculiformis	14	1–3 m	Volcanic	Krasnozem	120–160	24°	Westerly	Grassland with few shrubs
P25	Tai Po Kau (082817)	A. confusa	25	1.5 m	Volcanic	Krasnozem	380–420	19°	Northwesterly	Grassland with few shrubs
P35	Tai Lam (965785)	A. confusa	35	1.5 m	Granite	Red-yellow podzol	260–290	15°	Easterly	Grassland with few shrubs

Notes:

1. * The ages and tree spacing of the plantations were provided by the Agriculture and Fisheries Department.

2. No fire events were recorded for all the sites after tree planting.

3. NA: not available.

secondary forest of *Pinus massoniana* (Table 1). The site was converted to a borrow area from which fill materials were excavated up to a depth of 8 m. The surface was then graded to 25° slopes and covered with 50 cm of overburden materials of semi-decomposed granite before revegetation with a mixture of *A. auriculiformis*, *A. confusa* and *L. confertus* in 1994 (NT/ North West Development Office, 1988). This is a severely disturbed site where the topsoil has been removed and replaced with a skeletal soil that is strongly acidic in reaction and deficient in nitrogen, phosphorus and cation nutrients (Tsang, 1997). To prevent surface erosion, herringbone drains were installed to intercept runoff water from the slopes.

The second site (P6) is located in the Tai Lam Country Park, also in northwest New Territories. Prior to enhancement planting with *A. auriculiformis* to accelerate forest development in 1993, the site was dominated by grassland intermingled with isolated shrubs. It was a typical pyrogenic community maintained by frequent fires (Hodgkiss *et al.*, 1981). The dominant soil type is red-yellow podzol, a finely structured soil derived from fine-grained granite (Grant, 1960). It is strongly acidic in reaction and contains low to moderate levels of nitrogen, but it is deficient in phosphorus and cation nutrients (Chau and Lo, 1980).

The third site (P13) is located in the Wan Tsai Peninsula east of the New Territories, with a westerly aspect and an elevation of 30–60 m. It is similar to site 1 in geology, soil type and antecedent land use, except that the area is covered by 13-year old *A. confusa*. Granite outcrops are abundant in this borrow area, which is a result of erosion and poor workmanship during the grading of the slopes. Compared to the other sites, it is relatively isolated from the hinterland.

The fourth site (P14) is located at Lau Shui Heung, in the Pat Sin Leng Country Park northeast of the New Territories. It is covered by 14-year old *A. auriculiformis*, which was planted at 1–3 m apart. Krasnosem derived from volcanic rocks sits on slopes with an average gradient of 24°. As fires were abundant before enhancement planting, the site was previously dominated by a mixture of grasses and shrubs.

The fifth site (P25) lies at an altitude of 380–420 m, in the Tai Po Kau Nature Reserve east of the New Territories. It is the largest and best preserved nature reserve in Hong Kong, but it was almost completely felled by the Japanese during World War II (Nicholson, 1996). Occasional fires from the 1950s to the early 1970s reduced the area into grassland intermingled with isolated shrubs. It is now covered by 25-year old *A. confusa*, which was planted at 1.5 m apart. Volcanic rocks have weathered to form krasnosem on the moderate slopes.

The sixth (P35) site is located in Tai Lam Country Park, west of the New Territories. At a close proximity to the second site, it is similar in aspect, geology, soil type and antecedent land use, except that it is planted with 35-year old *A. confusa*. Immediately after World War II, there was an influx of refugees to Hong Kong which necessitated construction of the Tai Lam Chung Reservoir to meet the demand for water supply. Because of this, it represents one of the oldest sites that was reforested to protect the catchment for the water supply.

According to Aber's (1993) classification of land disturbances, sites P5 and P13 suffered from a severe disturbance of soil destruction in which the topsoils were removed. Conversely, sites P6, P14, P25 and P35 suffered from a mild disturbance of only the vegetation. It is easier to restore the vegetation on a mildly disturbed site than it is on a severely disturbed site.

Methods

A vegetation survey was conducted at the six study sites from March to early July 1999. In each of the study sites, four 10×10 m^2 plots were demarcated for the measurement of vegetation. A sampling area of 400 m^2 is considered adequate regarding the size and patchy distribution of woodland plantations in Hong Kong (Zhuang, 1993). In each plot, the height, basal diameter, diameter at breast height (dbh) and crown cover of the reforested species were recorded. Both living and dead trees were counted. A nested 5×5 m^2 sub-plot was earmarked at a corner of each 10×10 m^2 plot for the floristic inventory survey of the understorey layer. All trees, shrubs and vines were identified but not the herbs, ferns and grasses. Tree height was either measured directly (< 2 m) or indirectly (> 2 m) using trigonometric principles. The basal diameter and dbh were measured by use of a caliper or a measuring tape. For trees characterized by coppice growth, the dbh of all the stems were measured and summed up to give the required value. The tree crown cover was estimated by a crown-diameter method (Mueller-Dombois and Ellenberg, 1974).

Data obtained from each of the four 10×10 m^2 plots of a particular site were pooled in the statistical analysis to give the mean values and standard deviations. To facilitate comparison between sites, the density of the overstorey and understorey tree species was converted to the number of individuals per hectare. The number of understorey species in each site was tabulated according to family, genus and species. Likewise, the number of species was broken down into the different growth forms of tree, small tree, shrub, sub-shrub and vine. The Shannon indices representing species' diversity were calculated for each site.

Results

Overstorey stand characteristics

Each site is dominated by an overstorey layer of revegetated species, being more homogeneous in the younger plantations than in their older counterparts. The canopy has closed up for the 13-, 14-, 25- and 35-year-old plantations but not for their 5- and 6-year-old counterparts. Tree density varied considerably among the sites, ranging from 2,150 nos./ha to 7,625 nos./ha (Table 2). The highest density coincided with the 5-year-old polyculture stand of *A. auriculiformis*, *A. confusa* and *L. confertus* (P5), where no dead trees were recorded. Die-back of the reforested species occurred in the rest of the sites, ranging from 75 nos./ha for P6, P14 and P25 to 350 nos./ha for P35. While there is no discernible pattern of die-back with the age of the stands, it tends to be higher for *A. confusa* than for *A. auriculiformis*, resulting in a relatively low tree density in the P13 (2,150 nos./ha) and P35 (2,700 nos./ha) stands.

Average tree height followed the order of P14 (8.71 m) > P35 (8.22 m) > P13 (6.10 m) > P6, P25 (5.90 m) > P5 (3.41 m). There are more trees <2 m in height in P5 than in any other sites, although a large proportion of this is constituted by *A. confusa* and *L. confertus* (Table 2). Conversely, 98% of the *A. auriculiformis* in the same polyculture stand yields a height greater than 2 m, being comparable to that of the same species in P6. Among the monoculture stands, 78% of the 13-year-old *A. confusa* is greater than 2m in height compared to 100% of the 25-year-old stand. Therefore, tree height varied with species and age of the plantations. Overall, *A. auriculiformis* outperformed *A. confusa* and *L. confertus* in height growth within and between sites. For instance, the average height of *A. auriculiformis* (5.44 m) is more than double that of *A. confusa* (2.18 m) and *L. confertus* (2.49 m) in the 5-year-old polyculture stand. It increases progressively to 5.90 m and 8.71 m in the 6- and 14-year-old plantations, respectively. The height growth of *A. confusa* also increased with the age

Table 2. Plantation stand characteristics

| Site | Species | Nos. of trees per hectare | | | | | Mean height (m) | Mean basal diameter (mm) | Mean d.b.h. (cm) | Mean crown cover (m²) |
		Living*	Height <2m	Height >2m	Dead	Total				
P5	A. auriculiformis	2,625	50	2,575	0	2,625	5.44 (1.51)	9.94 (2.83)	8.24 (3.68)	4.62 (3.16)
	A. confusa	2,375	1,225	1,150	0	2,375	2.18 (1.23)	4.96 (2.15)	3.74 (2.65)	1.62 (1.09)
	L. confertus	2,625	1,150	1,475	0	2,625	2.49 (1.32)	2.93 (1.64)	2.07 (0.95)	0.96 (0.72)
	Total	7,625	2,425	5,200	0	7,625	3.41 (2.01)	5.87 (3.64)	5.49 (4.03)	2.42 (2.56)
P6	A. auriculiformis	5,550	50	5,500	75	5,625	5.90 (2.03)	7.30 (2.55)	6.15 (2.77)	2.62 (1.64)
P13	A. confusa	2,150	475	1,675	275	2,425	6.10 (3.90)	15.13 (9.98)	21.64 (11.65)	13.61 (12.58)
P14	A. auriculiformis	2,625	0	2,625	75	2,700	8.71 (2.38)	13.39 (4.28)	12.31 (5.35)	8.04 (5.69)
P25	A. confusa	4,750	0	4,750	125	4,875	5.90 (1.81)	8.26 (3.10)	7.24 (3.50)	4.65 (2.98)
P35	A. confusa	2,700	25	2,675	350	3,050	8.22 (2.94)	19.09 (7.23)	14.93 (7.11)	13.19 (10.50)

Notes:
1. *The total number of living trees counted at each site is used as the base for taking all the mean and standard deviation values.
2. Values in parentheses represent standard deviation.

of the plantations, averaging 2.18 m, 6.10 m and 8.22 m in the 5-, 13- and 35-year-old plantations respectively. The only exception was found in the 25-year-old plantation where the average height barely reached 5.90 m.

The basal diameter, dbh and crown cover largely increased with the age of the plantations, except for the 5-year-old *A. auriculiformis* and the 25-year-old *A. confusa*.

Understorey floristic composition

A total of 57 native wild growing vascular species were found in the sites, belonging to 30 families and 45 genera (Table 3). Approximately 28% of these species were tree taxa, representing 4.1% of the total tree flora in Hong Kong.

The number of species recorded in the sites followed the order of P6 (32) > P25(30) > P35(24) > P14(21) > P13(9) > P5(7), suggesting that the understorey species, as well as the families and genera they represent, do not necessarily increase with the age of the plantations. The breakdown of the species into different growth forms (Table 4) is summarized below:

Number of trees/small trees:
P6(8), P25(8), P14(8) > P35(5) > P13(1) > P5(1);
Number of shrubs/sub-shrubs:
P6(15) > P35(11) > P25(10) > P14(7) > P13(5) > P5(2);
Number of vines:
P25(12) > P6(9) > P35(8) > P14(6) > P5(4) > P13(3)

The 6-year-old *A. auriculiformis* plantation at Tai Lam contained the greatest number of understorey tree and shrub species, while the 25-year-old *A. confusa* plantation at Tai Po Kau contained the most vine species. Parallel to this, the 5-year-old polyculture stand recorded only a few tree and shrub species.

Table 3. Numbers and index showing species' diversity of understorey

Site	P5	P6	P13	P14	P25	P35	Overall
No. of families	6	21	6	15	19	14	30
No. of genera	7	29	8	18	27	22	45
No. of species	7	32	9	21	30	24	57
Shannon index	1.89	2.47	1.84	2.25	2.86	2.36	

A slightly different pattern is observed for the actual occurrences of individual species on each of the 100-m^2 areas. There seem to be more trees and more small trees in the older plantations than in their younger counterparts (Table 4). For instance, P35 recorded a total of 108 individuals compared to 80 for P14, 47 for P6, 38 for P25, 3 for P13 and 1 for P5.

None of the 57 species was found to occur at all the sites (Table 5). Only four species were found to occur at five of the study sites, namely *Embelia laeta, Glochidion wrightii, Psychotria serpens* and *Rhaphiolepis indica.* Similarly, eight species occurred at four of the sites, including *Breynia fruticosa, Evodia lepta, Gnetum montanum, Litsea rotundifolia Hemsl. var. oblongifolia, Psychotria rubra, Rhodomyrtus tomentosa, Smilax china* and *Smilax glabra.* Twenty-six species (45.6% total) were found to occur at only one of the sites. These include *Rhus succedanea* (P5), *Baeckea frutescens, Berchemia racemosa, Hedyotis acutangula, Helicteres angustifolia, Millettia sp., Pluchea indica, Rhus hypoleuca, Verbenaceae sp., Zanthoxylum avicennae* (P6), *Maesa perlarius, Sageretia theezans, Viburnum odoratissimum* (P13), *Cratoxylum cochinchinense, Gardenia jasminoides, Ilex pubescens, Itea chinensis, Rourea microphylla* (P14), *Columella corniculata, Graphistemma pictum, Liquidambar formosana, Mussaenda pubescens, Persea chekiangensis, Rubus leucanthus* (P25), *Ardisia crenata, Tetracera asiatica* and *Zanthoxylum nitidum* (P35).

Table 4. Number of species by growth form

Growth form	P5 N	P5 n	P6 N	P6 n	P13 N	P13 n	P14 N	P14 n	P25 N	P25 n	P35 N	P35 n	Overall N
Tree	0	0	3	10	0	0	0	0	5	33	2	11	5
Small tree	1	1	5	37	1	3	8	80	3	5	3	97	11
Shrub	2	3	13	481	5	33	7	121	8	82	10	115	19
Sub-shrub	0	0	2	16	0	0	0	0	2	15	1	3	3
Vine	4	7	9	141	3	10	6	17	12	161	8	30	19
Others*		2		1		0		0		3		0	
Total	7	13	32	686	9	46	21	218	30	299	24	256	57

Notes:

1. N, number of species; n, occurrence of individuals.

2. * 'Others' include undetermined plant individuals, mostly woody seedlings.

Table 5. Understorey woody species

Species	Family	Growth form*	No. of sites occurred	P5	P6	P13	P14	P25	P35
Acronychia pedunculata	Rutaceae	ST	2	—	—	—	2	1	—
Adinandra millettii	Theaceae	ST	3	—	7	—	36	—	2
Aporusa dioica	Euphorbiaceae	S	3	—	15	—	5	—	37
Archidendron lucidum	Mimosaceae	T	2	—	3	—	—	17	—
Ardisia crenata	Myrsinaceae	S	1	—	—	—	—	—	4
Baeckea frutescens	Myrtaceae	S	1	—	10	—	—	—	—
Berchemia racemosa	Rhamnaceae	V	1	—	1	—	—	—	—
Breynia fruticosa	Euphorbiaceae	S	4	—	1	3	—	7	3
Columella corniculata	Vitaceae	V	1	—	—	—	—	32	—
Clerodendrum fortunatum	Verbenaceae	S	1	—	18	—	—	—	—
Cratoxylum cochinchinense	Hypericaceae	ST	1	—	—	—	1	—	—
Embelia laeta	Myrsinaceae	V	5	—	63	1	4	13	11
Eurya japonica	Theaceae	S	3	—	5	—	—	2	11
Evodia lepta	Rutaceae	ST	4	—	1	—	1	3	9
Ficus variolosa	Moraceae	ST	2	—	2	—	2	—	—
Gardenia jasminoides	Rubiaceae	S	1	—	—	—	5	—	—
Glochidion wrightii	Euphorbiaceae	S	5	—	5	1	1	7	2
Gnetum montanum	Gnetaceae	V	4	2	1	—	—	16	2
Graphistemma pictum	Asclepiadaceae	V	1	—	—	—	—	2	—
Hedyotis acutangula	Rubiaceae	SS	1	—	13	—	—	—	—
Helicteres angustifolia	Sterculiaceae	S	1	—	59	—	—	—	—
Ilex asprella	Aquifoliaceae	S	2	—	15	—	—	—	36
Ilex pubescens	Aquifoliaceae	ST	1	—	—	—	7	—	—
Itea chinensis	Escalloniaceae	ST	1	—	—	—	17	—	—
Liquidambar formosana	Hamamelidaceae	T	1	—	—	—	—	2	—
Litsea rotundifolia Hemsl. var. *oblongifolia*	Lauraceae	ST	4	—	26	—	14	1	86
Maesa perlarius	Myrsinaceae	S	1	—	—	3	—	—	—
Melastoma candidum	Melastomataceae	S	2	1	9	—	—	—	—
Melastoma dodecandrum	Melastomataceae	SS	2	—	3	—	—	5	—
Melastoma sanguineum	Melastomataceae	S	2	—	—	12	—	3	—
Millettia sp.	Papilionaceae	V	1	—	26	—	—	—	—
Morinda parvifolia	Rubiaceae	V	3	1	24	—	—	—	4
Mussaenda pubescens	Rubiaceae	V	1	—	—	—	—	2	—
Persea chekiangensis	Lauraceae	T	1	—	—	—	—	12	—
Persea polyneura	Lauraceae	T	3	—	5	—	—	1	3
Pluchea indica	Compositae	S	1	—	23	—	—	—	—
Psychotria rubra	Rubiaceae	S	4	—	—	14	12	2	5
Psychotria serpens	Rubiaceae	V	5	2	—	7	1	2	1
Rhaphiolepis indica	Rosaceae	S	5	2	87	—	24	5	10
Rhodomyrtus tomentosa	Myrtaceae	S	4	—	231	—	72	50	4

Table 5. (Cont'd)

Species	Family	Growth form*	No. of sites occurred	P5	P6	P13	P14	P25	P35
Rhus hypoleuca	Anacardiaceae	ST	1	—	1	—	—	—	—
Rhus succedanea	Anacardiaceae	ST	1	1	—	—	—	—	—
Rourea microphylla	Connaraceae	V	1	—	—	—	1	—	—
Rubus leucanthus	Rosaceae	V	1	—	—	—	—	3	—
Rubus reflexus	Rosaceae	V	3	—	14	—	—	19	7
Sageretia theezans	Rhamnaceae	V	1	—	—	2	—	—	—
Sarcandra glabra	Chloranthaceae	SS	2	—	—	—	—	10	3
Schefflera octophylla	Araliaceae	T	3	—	2	—	—	1	8
Smilax china	Smilacaceae	V	4	—	9	—	6	15	2
Smilax corbularia	Smilacaceae	V	2	—	1	—	—	19	—
Smilax glabra	Smilacaceae	V	4	2	2	—	3	35	—
Smilax lanceifolia	Smilacaceae	V	2	—	—	—	2	3	—
Tetracera asiatica	Dilleniaceae	V	1	—	—	—	—	—	1
Viburnum odoratissimum	Caprifoliaceae	ST	1	—	—	3	—	—	—
Viburnum sempervirens	Caprifoliaceae	S	2	—	—	—	2	6	—
Zanthoxylum avicennae	Rutaceae	S	2	—	3	—	—	—	3
Zanthoxylum nitidum	Rutaceae	V	1	—	—	—	—	—	2

* T = Tree; ST = Small tree; S = Shrub; SS = Sub-shrub; V = Vine.

Discussion

Plantation stand characteristics

In Hong Kong the stand characteristics of tree plantations are rarely studied, not to mention their effect on the understorey. Tree density varied from 2,150 nos./ha to 7,625 nos./ha in the plantations primarily due to different silvicultural practices and the die-back of the species. With few exceptions tree density tends to decrease with the age of the plantations. According to information supplied by the Agriculture and Fisheries Department, the spacing of planting varied from 1 m to 3 m, although most of the sites were planted at a spacing of 1.5 m. As trees are established, they are subjected to thinning, the effects of wind and attack by pathogens. The shallow-rooted *A. auriculiformis* is easily blown over by the wind, a drawback which limits its use in the coastal region (Evans, 1992). *A. confusa* is more vulnerable to attack by pathogens than *A. auriculiformis*, resulting in higher die-back in sites P13 and P35. This is expected because *A. confusa* was introduced to Hong Kong much earlier than *A. auriculiformis* and the first generation of the introduced species is relatively immune to pest attack (Evans, 1992). The variable tree density indirectly affects the establishment of the understorey native species through its direct effect on the light intensity of the forest floor (MacDougall and Kellman, 1992; Messier *et al.*, 1998).

A. auriculiformis outperformed *A. confusa* and *L. confertus* in height growth within and between sites. Tree height tends to increase with the age of the plantations, except at P25 where the average height of *A. confusa* is less than that at P13. With a northwesterly aspect and an elevation of 380–420 m, highest among the sites, the growth performance of this tropical species is likely affected by the cooler environment. There are two implications related to the differential height growth of the acacias. First, the selection of the species mix in rehabilitation planting should be carefully conceived, otherwise, the slower growing species (e.g. *A. confusa*) can easily be dwarfed by its faster growing counterparts (e.g. *A. auriculiformis* and *L. confertus*), hence undermining its ecological role as a nitrogen fixer. Second, consideration must be given to the altitudinal limit of *A. confusa* which is not cold-tolerant (Chinese Tree Species Editorial Committee, 1976).

Understorey floristic composition

A total of 57 species, representing 45 genera and 30 families, were recorded at the six plantations. In a separate local study on ten plantation plots aged 15–50 years, Zhuang and Yau (1998) found a total of 184 species, belonging to 139 genera and 70 families. However, it is premature to conclude that the sites under investigation are less diverse in their understorey floristic composition because ferns, herbs, grasses, and sedges were excluded in the present survey.

Species' richness does not increase with the age of the plantations, as indicated by the variable Shannon indices of 1.84–2.86. This finding differs from the study of 10 plantations aged 15-50 years by Zhuang and Yau (1998). The occurrence of the 57 species is probably constrained by site-specific factors resulting in diverse genera and the families they represent. None of the 57 species was found to occur in all the sites. This is expected because the sites transcend a plantation history of 5 to 35 years during which the forest environment, such as light intensity, edaphic conditions, microclimate, could have been modified to different extents (Chan and Thrower, 1988).

Twelve out of the 57 species occurred at four to five sites, suggesting that they can adapt to a wide range of environments, although the abundance of individual species is closely related to the age of the plantations. For instance, the occurrence of *Breynia fruiticosa*, *Evodia lepta* and *Smilax glabra* tends to increase with the age of the plantations, while *Rhaphiolepis indica* and *Rhodomyrtus tomentosa* show a reverse trend. As a matter of fact, *R. indica* and *R. tomentosa* are shade-intolerant and will gradually fade out in a forested environment (Yu and Pie, 1962).

Twenty-six species (45.6% total) are found to occur only at one of the sites, suggesting a narrow tolerance level of these species to the environment. Among these species, *Liquidambar formosana* and *Schefflera octophylla* are mid-successional species in the local environment (Chang *et al.*, 1989; Lay *et al.*, 1999). This clearly indicates that some forms of ecological succession have already begun in the older plantations and in this connection, 44 occurrences of trees (*Archidendron lucidum*, *L. formosana*, *Persea chekiangensis*, *P. polyneura* and *S. octophylla*) were recorded in P25 and P35 compared to only 10 occurrences (*A. lucidum*, *P. polyneura* and *S. octophylla*) in the remaining sites. Understorey development is therefore highly variable among these exotic plantations. The occurrence and abundance of tree species clearly increased with the age of the plantations, but not the overall species' richness.

Borrow area restoration

The understorey layer is poorly developed underneath the 5- and 13-year-old plantations in the borrow areas. There are only 7–9 species, compared to 21–32 species recorded for the other sites. Of the species present, they are mostly vines and shrubs, while tree species are lacking. The low biodiversity is a cause of concern because unlike the studies by Lugo (1988) and Geldenhuys (1997), the tree plantations seem to have no catalytic effect on the establishment of the forest species. Borrow area restoration begins with the grading of the surface and the laying of a 50-cm overburden dominated by semi-decomposed parent materials. This substrate is deficient in nutrients and deprived of seed reserves and other vegetation materials, typical of the most severely degraded landscape (Aber, 1993). It is then hydroseeded with the introduced grasses of *Cynodon dactylon*, *Lolium perenne* and *Paspalum notadum*, in mixture or singly, before pit planting with *A. confusa* and other species. These exotic but fast-growing species can provide a rapid vegetative cover to protect the soil and stabilize the newly-engineered slopes (Geotechnical Control Office, 1984), although their suitability for ecosystem restoration has not been ascertained in the local environment (Tsang, 1997). The hydroseeded grasses faded out gradually with the canopy closure of the plantation, resulting in the development of an even-aged stand that is simple in structure and has limited perches and food resources to attract birds. Indeed, seeds of *A. confusa* and *A. auriculiformis* possess a hard pod that is unattractive to frugivorous birds, resulting in limited seed dispersal. This results in poor development of the understorey layer, as compared to the grassy slopes of sites P6, P14, P25 and P35 where there was an ample supply of genetic materials prior to the enhancement planting. In addition, the low nitrogen content, strong acidity, and low moisture availability of the borrow area soils (Tsang, 1997) and poor management practices, such as weeding (Zhuang, 1997), can also inhibit understorey tree establishment. Hence, exotic tree plantations are not necessarily appropriate to foster ecosystems for the rehabilitation of severely damaged sites if natural stands are not available within a reasonable distance. The 13-year-old plantation at Wan Tsai peninsula suffers from this drawback because the site is relatively isolated from the hinterland. The polyculture of exotic and native species is therefore preferred to the monoculture of exotic species, a strategy also advocated by Lugo (1988, 1997). It is necessary to transport as many genetic materials as possible to the site so that the tougher species will survive the harsh environment, induce changes in site conditions and pave the way for the establishment of native tree species. As legumes play an important role in augmenting the soil nitrogen supply, their capacity to fix atmospheric nitrogen can be adversely affected by the strong acidity of the soil (Reddell, 1993). A more vigorous approach is needed to correct the soil pH before planting. Nature will then look after itself, resulting in the restoration of native species in the shortest possible time, otherwise a species-poor community will develop as in sites P5 and P13.

The establishment of understorey species in the borrow area can also be optimized if part of the topsoil is stockpiled for use as a growth substrate to provide the necessary propagules. According to our observation in the field, the fruit-bearing shrubs of *Melastoma candidum*, *Melastoma sanguineum*, *Phyllanthus emblica*, *Rhodomyrtus tomentosa* and *Viburnum odoratissimum* are abundant in the nearby undisturbed slopes. Had the topsoil been preserved, the regeneration of these species would have been possible either from the seeds or the rootstocks. More research is needed to investigate the planting strategy of this scenario because shading of the fruit-bearing shrubs by the overstorey species will undermine their potential as an attractant of birds.

Role of enhancement planting on grassy slopes

In Hong Kong enhancement planting of grassy slopes has been implemented since the 1960s, with the dual objectives of containing the spread of hill fires and accelerating native forest development. Plantations are believed to have a nurse effect which promotes the invasion of native species and attracts seed dispersers (Lugo, 1997). While frequent fires will arrest ecological succession at the grassland stage (Daley, 1975; Thrower, 1984), shrubs will invade the grassland when fire has been absent for 5 to 6 years (Hodgkiss *et al.*, 1981). In the absence of fires for around 20 years, the degraded sites could potentially be restored with tree species typical of the secondary native forest which will close up in another 10–20 years time (Zhuang, 1997). An obvious question then arises: is it worthwhile to reforest sites P6, P14, P25 and P35 with exotic legumes?

Unlike the borrow area where propagules are limited in the overburden layer, the grassy slopes contain a substantial amount of plant materials which will be re-established with removal of the arresting factor. In Hong Kong the typical grassy slopes contain 14 (Yau, 1996) to more than 31 native species (Thrower and Thrower, 1986) depending on the frequency of fire. As mentioned earlier, the number of understorey woody species amounted to 32, 21, 30 and 24 in sites P6, P14, P25 and P35 respectively, a large proportion of which is constituted by shrubs and vines. The corresponding stocking density of understorey native small trees and trees amounted to 4,700, 8,000, 3,700, and 10,800 nos./ha which may readily replace the exotic legumes. Research on management prescriptions for these plantations is urgently required to foster growth of the desirable species into a self-sustaining secondary forest. However, it is premature to conclude that enhancement planting of exotic species is the only option to accelerate the restoration of native forests because protection of the derelict sites against fire and other disturbances may produce similar results with less cost and effort. If native tree species are used in enhancement planting, it will truly enhance biodiversity and accelerate forest establishment, as illustrated by the latest woodland compensatory project of the new airport at Tung Chung in which 50 native species were planted with considerable success on similar grassy slopes (Lay *et al.*, 1999). The use of exotic legumes in enhancement planting needs to be reviewed as the *in situ* soils already contain a substantial amount of nitrogen to support the growth of more nutrient-demanding tree species (Marafa and Chau, 1999). Therefore, the ecological role of legumes to augment soil nitrogen is less critical on the grassy slopes than it is in the nitrogen-poor borrow area. It is not known if understoreys in sites P6, P14, P25 and P35 have developed by a process of facilitation (Purdie and Slayter, 1976) or whether they might have developed as rapidly, or even more rapidly, without the plantation. Perhaps the inclusion of control sites will provide an answer to this question.

References

Aber, J. D. "Restored Forests and the Identification of Critical Factors in Species-site Interactions." In *Restoration Ecology: A Synthetic Approach to Ecological Research,* edited by Jordan III, W. R., Gilpin, M. E. and Aber, J. D., pp. 241–250. New York: Cambridge University Press, 1993.

Agriculture and Fisheries Department. *Agriculture and Fisheries Department Annual Reports, 1980–1995.* Hong Kong: Government Printer.

Agriculture and Fisheries Department. *Progress Reports 1985–1988 of Hong Kong.* Paper prepared by the 13th Commonwealth Forestry Conference 1989.

Bradshaw, A. D. "The Reclamation of Derelict Land and the Ecology of Eco-systems." In *Restoration Ecology: A Synthetic Approach to Ecological Research,* edited by Jordan III, W. R., Gilpin, M. E., Aber, J. D., pp. 53–74. New York: Cambridge University Press, 1993.

Chan, Y. C. and Thrower, L. B. "Succession Taking Place under *Pinus Massoniana.*" In *Memoirs of the Hong Kong Natural History Society,* edited by Hodgkiss, I. J., pp. 59–65. Hong Kong: Hong Kong Natural History Society, 1988.

Chang, H. D., Wang, B. S., Hu, Y. K., Bi, P. X., Chung, Y. H., Lu, Y. and Yu, S. X. "Vegetation of Hong Kong." *Acta Scientiarum Naturalium Universityatis Sun Yat Sen* 8(2), 1989, 172 pp. (in Chinese).

Chau, K. C. and Lo, W. K. "The Pinus Scrub Community as an Indicator of Soils in Hong Kong." *Plant and Soil* 56, 1980, pp. 243–254.

Chinese Tree Species Editorial Committee. *The Afforestation Techniques of Major Tree Species in China, Vol. 1.* Beijing: Agriculture Publishing Institute, 1976 (in Chinese).

Daley, P. A. "Man's Influence of the Vegetation of Hong Kong." In *The Vegetation of Hong Kong: Its Structure and Change: Proceedings of a Weekend Symposium of the Royal Asiatic Society (Hong Kong Branch),* edited by Thrower, L. B., pp. 44–56. Hong Kong: Royal Asiatic Society (Hong Kong Branch), 1975.

Dudgeon, D. and Corlett, R. *Hills and Streams: An Ecology of Hong Kong.* Hong Kong: Hong Kong University Press, 1994.

Evans, J. *Plantation Forestry in the Tropics: Tree Planting for Industrial, Social, Environmental, and Agroforestry Purposes*, 2nd ed. Oxford: Clarendon Press, 1992.

Geldenhuys, C. J. "Native Forest Regeneration in Pine and Eucalypt Plantations in Northern Province, South Africa." *Forest Ecology and Management* 99(1–2), 1997, pp. 101–115.

Geotechnical Control Office. *Geotechnical Manual for Slopes*, 2nd ed. Hong Kong: Government Printer, 1984.

Giller, K. E. and Wilson, K. *Nitrogen Fixation in Tropical Cropping Systems.* Wallingford: C. A. B. International, 1991.

Grant, C. J. *The Soils and Agriculture of Hong Kong.* Hong Kong: Government Printer, 1960.

Hodgkiss, I. J., Thrower, S. L. and Man, S. H. *An Introduction to Ecology of Hong Kong, Vol. 1.* Hong Kong: Federal, 1981.

Kendle, A. D. and Bradshaw, A. D. "The Role of Soil Nitrogen in the Growth of Trees on Derelict Land." *Arboricultural Journal* 16, 1992, pp. 103–122.

Lamb, D. "Large-scale Ecological Restoration of Degraded Tropical Forest Lands: The Potential Role of Timber Plantations." *Restoration Ecology* 6(3), 1998, pp. 271–279.

Lay, C. C., Chan, S. K. F. and Chan, Y. K. "The Utilization of Native Species in Local Afforestation Programme." *Occasional Paper.* Hong Kong: Agriculture and Fisheries Department, 1999.

Lugo, A. E. "The Future of the Forest: Ecosystem Rehabilitation in the Tropics." *Environment* 30(7), 1988, pp. 17–20, 41–45.

Lugo, A. E. "The Apparent Paradox of Reestablishing Species Richness on Degraded Lands with Tree Monocultures." *Forest Ecology and Management* 99 (1–2), 1997, pp. 9–19.

MacDougall, A. and Kellman, M. "The Understorey Light Regime and Patterns of Tree Seedlings in Tropical Riparian Forest Patches." *Journal of Biogeography* 19, 1992, pp. 667–675.

Marafa, L. M. and Chau, K. C. "Morphological and Chemical Properties of Soils along a Vegetation Gradient Affected by Fire in Hong Kong." *Soil Science* 164, 1999, pp. 683–691.

Messier, C., Parent, S. and Bergeron, Y. "Effects of Overstorey and Understorey Vegetation on the Understorey Light Environment in Mixed Boreal Forests." *Journal of Vegetation Science* 9, 1998, pp. 511–520.

Mueller-Dombois, D. and Ellenberg, H. *Aims and Methods of Vegetation Ecology.* New York: Wiley, 1974.

Nicholson, B. "Tai Po Kau Nature Reserve, New Territories, Hong Kong: A

Reafforestation History." *Asian Journal of Environmental Management* 4(2), 1996, pp. 103–119.

NT/North West Development Office . *Tai Tong East Borrow Area: Outline Layout and Landscape Plan.* Hong Kong: New Territory Development Department, 1988.

Oliver, C. D. and Larson, B. C. *Forest Stand Dynamics.* New York: Wiley, 1996.

Parrotta, J. A. "Secondary Forest Regeneration on Degraded Tropical Lands: The Role of Plantations as 'Foster Ecosystem'." In *Restoration of Tropical Forest Ecosystems: Proceedings of the Symposium held on October 7–10, 1991,* edited by Lieth, H. and Lohmann, M., pp. 63–73. Boston: Kluwer Academic, 1993.

Purdie, R. W. and Slayter, R. O. "Vegetation Succession after Fire in Sclerophyll Woodland Communities in Southeastern Australia." *Australian Journal of Ecology* 1, 1976, pp. 223–236.

Reddell, P. "Soil constraints to the Growth of Nitrogen-fixing Trees in Tropical Environments." In *Symbioses in Nitrogen-Fixing Trees,* edited by Subba Rao, N. S. and Rodriguez-Barrueco, C., pp. 65–83. New York: International Science Publishers, 1993.

Thrower, L. B. "The Vegetation of Hong Kong." In *The Vegetation of Hong Kong: Its Structure and Change: Proceedings of a Weekend Symposium of the Royal Asiatic Society (Hong Kong Branch),* edited by Thrower, L. B., pp. 21–43. Hong Kong: Royal Asiatic Society (Hong Kong Branch), 1975.

Thrower, S. L. *Hong Kong Country Parks.* Hong Kong: Government Printer, 1984.

Thrower, S. L. and Thrower, L. B. "The Effect of Periodic Fires on the Development of a Hill-side Community in Hong Kong." In *Memoirs of the Hong Kong Natural History Society,* edited by Hodgkiss, I. J., pp. 45–49. Hong Kong: Hong Kong Natural History Society, 1986.

Tsang, P. *Early Ecosystem Restoration in Hong Kong: A Case Study of the Tai Tong East Borrow Area.* Unpublished M. Phil. Thesis. Hong Kong: The Chinese University of Hong Kong, 1997.

Yau, M. *Effect of Hill Fire on Soil and Vegetation in Tai Mo Shan Country Park, Hong Kong.* Unpublished M. Phil. Thesis. Hong Kong: The Chinese University of Hong Kong, 1996.

Yu, M. and Pie, C. "Effect of Vegetation on the Composition of Humus and Active Mineral Substances in Soil of Kwangtung and Kwangsi." *Acta Pedologica Sinica* 10(1), 1962, pp. 29–43 (in Chinese).

Zhuang, X. *Forest Succession in Hong Kong.* Unpublished Ph.D. Thesis. Hong Kong: University of Hong Kong, 1993.

Zhuang, X. "Rehabilitation and Development of Forest on Degraded Hills of Hong Kong." *Forest Ecology and Management* 99(1–2), 1997, pp. 197–201.

Zhuang, X. and Yau, M. "An Investigation of Floristic Diversity of Three Types of Plantations in Hong Kong." *Journal of Tropical and Subtropical Botany* 6(3), 1998, pp. 196–202 (in Chinese).

Towards Better Management and Governance of Soil and Tree Resources in Urban Hong Kong

C.Y. Jim
Department of Geography and Geology, The University of Hong Kong

Introduction

Trees and urbanism

Urbanization in human history denotes a major landmark in cultural advancement, allowing us to partly escape from the vicissitudes and vagaries of nature and to nurture a way of life that extends beyond the harsh and monotonous hand-to-mouth existence. The progressive intensification of urban development triggered especially after the onset of the Industrial Revolution, however, has engendered a host of ills and woes that have marred the ideal of city living (Mumford, 1961). Urbanism has brought about excessive detachment from nature, both in the physical sense as cities grow to enormous proportions leaving wilderness far away from settled areas, and in the psychological sense in that our innate urge to be in the company of nature can hardly be satisfied in the overwhelmingly artificial and stressful conditions typical of the urban milieu.

The research grant support provided by the Environment and Conservation Fund and the Woo Wheelock Green Fund is hereby gratefully acknowledged. I would like to extend my appreciation to the assistance provided by my colleagues and officers of various government departments in providing information and help in the field work.

Rather than being passively cut off from nature, since antiquity humans began to find ways to usher in and introduce natural elements into and around our dwellings. There are numerous forms of expression in different cultures and peoples at different times and places of this inborn human desire to reconstitute and strengthen the tenuous links with mother nature. From the early days of urban development, the aristocracy tended to keep hunting grounds and other forms of semi-wild lands within their estates, together with neatly configured landscape gardens as symbols of the human triumph over nature as well as of a tamed surrogate of nature. Some of these initially countryside lands with high-calibre natural ingredients were occluded within the city proper and subsequently became public parks and gardens.

The urban poor, meanwhile, had to eke out a living cramped into tenement blocks devoid of nature or environmental quality. Degradation in hygiene conditions and the spread of contagious diseases were some of the impetuses to improve the residential environment. The garden-city utopia (Howard, 1902) began to attract followers and practitioners. City planners gradually realized these excesses and started to introduce greenery into built-up areas in the form of communal vegetation along roads and in public greenspaces (Kelcey, 1978; Lawrence, 1988). Trees found their way into urbanized areas and acquired legitimacy in an essential urban infrastructure, to the extent that it is now almost routine to include green components into new urban areas or in rejuvenated old enclaves. The expansion of suburbia extended green living through the medium of home gardens. Green belts (Nicholson-Lord, 1987) were devised to contain the unbridled sprawl of cities into the surrounding countryside.

Trees in urban space

The notion of having trees in cities has gradually been accepted as a premier greenery component serving the more utilitarian environmental and ornamental functions. The higher-level psychological and emotional benefits of having good-quality greenery in living and working areas were subsequently also recognized (Jim, 1996a). Trees in human settlements, a natural partnership, have been revived if not resurrected. Fine amenity vegetation in cities has secured a transcultural, timeless and universal appeal. It is widely advocated as a key element in a livable city, and a green city has become the target if not the pride of many municipal authorities worldwide. Cooperation between government and private forces in many cities has jointly ushered successfully the generous spread of urban trees.

The implementation of a green city programme is not without its obstacles (Schmid, 1975; Bradshaw et al., 1995). The urban habitat is not particularly conducive to the growth of plants and their associated wildlife alliances. The above- and below-ground constraints to tree expansion are both ubiquitous and daunting. There are also the less tangible institutional barriers that may pose additional restrictions and are often more difficult to overcome than physical and physiological restrictions (Grey and Deneke, 1986; Miller, 1997). In densely-developed areas and rapidly expanding cities, the tree-planting tradition may sometimes be submerged by the relentless urge to maximize land use. To ensure that this public good is not unduly pushed aside, the government has to play the roles of both steward and umpire. In the process of town planning, every effort should be made to increase plantable spaces and to upgrade the cityscape through planting. Urban foresters thus serve as the frontline professionals to fulfil the onus of the custodians of the urban forest and to enhance the ecological quality of the city (Platt et al., 1994).

Study objectives and study area

This study attempts to make a synoptic assessment of the current status of the urban forest in Hong Kong, with a view to surveying the multiple constraints on tree growth, including those in the subaerial environment as well as the soil envelope. The existing practices of tree care and management will also be evaluated to see if changes or improvements may be recommended (Grey, 1996). The fundamental questions that will be addressed are: what has been done and what could have been done? As a yardstick to gauge Hong Kong's performance, the wide panorama of urban forestry and arboricultural knowledge and practices elsewhere will be considered. Relevant state-of-art developments in the science of tree work will be evaluated for their applicability to the local scene. The possibility of conducting research with reference to the local-regional tree species and the urban conditions for tree growth will also be studied. It is hoped that the gap between science and policy can be closed.

The study area encompasses the core and relatively older urban areas on Hong Kong Island and Kowloon, situated mainly around Victoria Harbour. Development began as early as the 1840s from a small incipient township, and subsequently spread out along the northern shores of Hong Kong Island and sprawled across Kowloon to become a metropolis marked by an exceptionally high-rise and high-density urban morphology. The study area covers 124 km^2 out of a total territory area of 1095 km^2. It accommodates a population of 3.3 million out of a total of 6.9 million. To illustrate the inordinately high density of development, the average density reaches 27,000 persons/km^2, with the maximum district density rising to 54,000 persons/km^2, and the maximum spot density reaching 116,500 persons/km^2 (Census and Statistics Department, 2000).

The current status of the urban forest

Compact and cramped urban morphology

The rugged topography dominated by steep hills necessitated the leveling of the slopes into terraces and reclamation from the sea to create developable lands. These two modes of urban growth were initiated soon after the city was founded, and continue to date both in the main urban core and in new towns (Jim, 1989a). As such land production operations are both time consuming and expensive, there is an urge to develop a compact city right at the inception of urban growth. The rapid increase in the population has no doubt reinforced the need to go for high-density and high-rise development. The resultant town plan is extremely tightly packed in the sense that almost all land at the ground level is occupied either by roads or buildings, with little interstitial spaces for amenities and greenery (Jim, 1994a). This is the most fundamental limitation to any attempt to introduce greenery into the city.

Such a configuration tends to spread from the city centre to the peripheral areas, with only occasional deviations in some scattered neighbourhoods from the rather monotonous spread of buildings-cum-roads. In the several decades after WWII, the frantic development and redevelopment resulted in further increases in density in large stretches of the city. As a result, the territory has managed to squeeze its 6.9 million population largely within 175 km^2 (or about 16%) of its land areas. In the process, the urban environmental quality has been sacrificed, and plantable spaces either were not provided or were squeezed out. Extensive parts of the city are treeless and drab, and overall the tree stock is tiny in comparison with the population and the built masses (Jim, 1987a–b). The myth that a high-density city cannot accommodate much greenery, meanwhile, has become deeply ingrained. It is only in the last two decades or so that attempts have been made to reduce development density by decanting into new towns.

History of the urban forest

A survey of the large collection of historical photographs in Hong Kong (Hong Kong Museum of History, 1982) provides a visual record of the cityscape, including the location and condition of the urban greenery. Similarly, the old maps also register directly and indirectly the state of the town plan and by implication its built form. At the beginning of urban growth back in the mid-19th century, the town plan was relatively more open and porous, although pressures to make the best use of scarce developable land were already felt. There were widespread setbacks of buildings, and generous planting of trees at the roadsides and within the building grounds. Before the advent of motorized vehicles, which did not arrive until the end of the 19th century, trees were often planted on the edges of the carriageways outside the kerb line. Competition for physical space among buildings, vehicles and pedestrians was present but relatively mild.

The construction technology and the building code up until WWII basically encouraged the erection of three-to-four storey tenement blocks, resulting in extensive coverage of the city by rather monotonous low-rise structures. Development density gradually increased with the rapid growth in the population, resulting in an increase in site coverage and a tighter town plan. The post-WWII period witnessed a phenomenal jump in the population mainly due to massive immigration from mainland China, thus imposing additional pressures to raise development density in the first two post-war decades or so mainly by piecemeal redevelopment of old tenement blocks into high-rise buildings. The resultant land-use intensification as well as infilling resulted in a deterioration of the urban environment for both people and trees (Jim, 1998b, 1990a).

From the 1970s, the new town programme ushered in a new phase in urban growth, with the transfer of development to hitherto rural areas lying outside the main urban core. The opportunity to create new urban areas provided relief from the oppressive density of the old town. The new town plans in general are more porous and more amenable to planting, denoting a marked departure from the old haphazard ways of organic accretion.

Meanwhile, attempts were made to reduce the density in the old "metroplan" areas by renewal to a lower design population, and by decanting people into the new towns. Urban renewal should have provided many chances to open up the old town and introduce modern standards of landscape design. Unfortunately, in most renewal areas the process was too piecemeal and it occurred without an overall landscape plan, hence cityscape rejuvenation often failed to materialize (Jim, 1994a). It was only in the large-scale comprehensive redevelopment projects, and in new development areas on virgin lands acquired by hill terracing or reclamation, that some measures of an open town plan and tree insertion were implemented. Even in these limited cases of awakening, the greening was carried out without the benefit of a comprehensive city-wide landscape plan (Jim, 1998a).

Urban forest composition

The climate in Hong Kong is typical of the northern humid-subtropics, with hot-wet summers and cool-dry winters, interspersed by short intermediate periods. The monsoon weather system imposes major influences. The biogeographical situation may be characterized as an ecotone or transitional belt between two major biomes, namely humid-tropical and warm-temperate. The natural conditions are conducive to a rich floristic endowment which includes a good proportion of tropical species accompanied by some warm-temperate components. The pre-human vegetation cover is believed to be a semi-evergreen tropical forest of the monsoonal variety with a well-defined dry season, and hence the dominant evergreen trees are mixed with some deciduous members (Jim, 1986; Corlett, 1999).

The original forests have been quite thoroughly eradicated by centuries of human habitation and associated destructive activities. The early shifting cultivators were able to live largely in harmony with the forest. The subsequent sedentary farmers arriving from the north about 1000 years ago removed the forest cover. The loss of the primary forest, resulting in the demise of many plant and animal species, occurred in historical times. The remnant flora, however, is still relatively quite diversified. The small territory of Hong Kong harbours some 3000 vascular species, of which about 390 are trees (Jim, 1990b; Agriculture and Fisheries Department, 1993). The flora has been notably enriched, especially in the last 150 years, by plant introductions via intentional and inadvertent modes (Jim, 1991).

Many well-known tropical ornamental tree species have been brought into the city which operates as a free port and has many trade and personnel contacts with different parts of the world (Jim, 1992a; Zhuang et al., 1997).

Urban tree provenance

An analysis of the urban trees in the study area indicates an interesting range of history and provenance. Some are pre-urbanization relics in that they were the inherited remains of previous woodlands which in turn originated as products of hillslope afforestation. The encroachment of the city into the adjacent countryside occasionally allowed small pockets of woods or a few trees to stay in the future town plan (Jim, 1989a). This sort of salvaging resulted from deliberate actions to preserve the trees, or more often they were left by default in certain nooks and crannies. Invariably, such relics are found in the terraced hillslope areas where secondary woodlands tended to be more extensively nurtured, especially on Hong Kong Island. Often the species composition of the relic islands, mainly of the local woodland group, betrays their provenance.

Some urban trees were initially planted for one type of land use, and subsequently, due to changes in land use, shifts in lot boundaries or road alignments, they were trapped in different environs (Jim, 1992b). A common case in point is trees originally planted in private gardens, and then due to redevelopment or road upgrading, became trees at the roadside or in public open spaces. Some former military barracks surrounded by built-up areas and with generous tree cover have been designated for conversion into urban parks. The inherited mature trees can thus create almost an "instant" parkscape. The lands obtained by reclamation from the sea have no inherited greenery, and all trees there have to be nurtured from scratch in the usually harsh site conditions. In other built-up areas, tree are restricted to the grounds of government, institutional and community facilities, and the few low-density residential areas (Jim, 1987b, 1993b). Most urban trees are common tropical amenity species with a clear domination by exotics. The two major urban forests occur at the roadside and in parks.

Roadside trees

A recent census of all roadside trees in the study area (Jim, 1994b–g) indicates that only a limited number of roads is adorned with trees. The tree stock in this linear and stressful habitat amounts to about 25,000. Their distribution is extremely uneven, scattered on some 500 roads. They belong to 45 botanical families and 149 species, which is considered to be a high degree of diversity compared to other cities. Mainly broadleaf and evergreen, the trees are dominated by a number popular species, namely: *Aleurites moluccana* 12.9%, *Melaleuca quinquenervia* 7.54%, *Phoenix roebelenii* 7.0%, *Livistona chinensis* 7.0%, and *Caryota ochlandra* 5.3%. The roadside treescape is heavily represented by a small subset of common members, to the tune that the top 4 species take 34.4% of the tree population, the top 8 take 50%, and the top 14 take 66%. Most other species are uncommon to rare, and 20 species have a solitary specimen. As much as three-quarters of the trees have an exotic origin (Jim, 1996b–c).

At the roadside, the common botanical families and their principal constituent genera are summarized as follows:

- Arecaceae (Palmae, Palm Family) 17 species
  Archontophoenix, Caryota, Livistona, Phoenix, Roystonea, Washingtonia
- Moraceae (Mulberry Family) 12 species
  Artocarpus, Ficus, Morus
- Myrtaceae (Myrtle Family) 11 species
  Callistemon, Eucalyptus, Melaleuca, Syzygium
- Caesalpiniaceae (Senna Family) 10 species
  Bauhinia, Cassia, Delonix, Peltophorum
- Mimosaceae (Mimosa Family) 7 species
  Acacia, Albizia, Leucaena
- Papilionaceae (Bean Family) 7 species
  Dalbergia, Erythrina, Pterocarpus
- Lauraceae (Laurel Family) 7 species
  Cinnamomum, Litsea, Persea

Urban park trees

Ten major urban parks in the study area have trees with trunk diameters exceeding 15 cm, censused in a comprehensive study which was completed recently (Jim, 2000a–d). Altogether, some 11,000 semi-mature to mature trees were enumerated. The species' diversity, at 272 species derived from 65 botanical families, is surprisingly high. Exotic trees take up 73%, whereas exotic species 71%. Similar to the roadside situation, park trees are dominated by popular species. Most species are broadleaf evergreen, followed by broadleaf deciduous, and then palm. The most common species: *Acacia confusa* 10.7%, *Livistona chinensis* 10.1%, *Casuarina equisetifolia* .8%, and *Aleurites moluccana* 4.8%. The top 4 species take 30.4% of the tree population, the top 8 take 41.2%, and the top 14 take 53.1%. The majority of the species are either uncommon or rare, with as many as 118 species existing as solitary specimens confined to one venue. Many of the rare and solitary species are not planted outside the urban parks.

In the urban parks, the common botanical families and their principal constituent genera are summarized as follows:

- Arecaceae (Palmae, Palm Family) 32 species
  Archontophoenix, Caryota ,Chrysalidocarpus, Hyophorbe, Livistona, Neodypsis, Phoenix, Ptychosperma, Roystonea, Syagrus, Trachycarpus, Washingtonia
- Moraceae (Mulberry Family) 19 species
  Ficus, Morus
- Euphorbiaceas (Spurge Family) 18 families
  Aleurites, Bischofia, Bridelia, Macaranga, Mallotus, Sapium
- Caesalpiniaceae (Senna Family) 16 species
  Bauhinia, Cassia, Delonix, Peltophorum
- Myrtaceae (Myrtle Family) 16 species
  Callistemon, Eucalyptus, Melaleuca, Syzygium
- Lauraceae (Laurel Family) 11 species
  Cinnamomum, Litsea
- Mimosaceae (Mimosa Family) 9 species
  Acacia, Albizia, Leucaena
- Apocynaceae (Dogbane Family) 8 species
  Thevetia
- Meliaceae (Mahogany Family) 7 species
  Melia
- Papilionaceae (Bean Family) 7 species
  Erythrina, Pongamia, Pterocarpus

Trees in informal public greenspaces

The green and wooded slopes perching on the fringe of the city, particularly on Hong Kong Island, form a pleasant green backdrop to the metropolis. They exist as a kind of peri-urban woodland that defines the upslope boundary of urbanization and helps to circumscribe the urban sprawl. In a few locations, the woods extend downslope in a tongue-like configuration into the built-up areas, thus bringing their many environmental and ecological benefits into the city proper (Jim, 1990a). Such a juxtaposition of city and nature, literally furnishing the countryside right at the city's backyard, is a precious gift of Hong Kong, one that has been bequeathed largely by default rather than by design. The high ecological, landscape and amenity values bestowed by these natural areas (Flink and Searns, 1993), which cannot be emulated by any form of manicured urban park (Cole, 1986), deserve to be rigorously protected.

Unfortunately, the multiple values of these peri-urban woodlands have not been appropriately recognized both within and without the administration. Most of the public lands lie outside the city and the country park boundaries, and hardly receive any management and protection. Periodically, they are subject to intrusion by developments (Jim, 1989b). In general, they are treated as the city's land bank to be utilized for new buildings or road developments as required. The sense that this gem of the community should be designated as a green belt to contain further urban encroachment, or better still as a conservation area, thus far has not been adequately taken up. Proactive measures are needed to guard them from development pressures and to introduce woodland management to enhance their ecological ingredients (Jim, 1998a–b).

The city contains various forms of semi-natural or ruderal sites which are scattered in the form of tiny habitat islands embedded within the matrix of structures and roads. Some are incidental plots occurring as "left-over" pockets in odd locations not required for buildings, roads, or formal landscape treatment. These wild sites are the honeypots for the survival of some spontaneous species, mainly native but also some naturalized exotics, that can disperse and propagate of their own accord. A notable ruderal environment is the old stone walls, a vertical habitat which has been widely colonized by some trees mainly from the genus *Ficus* accompanied by a host of herbs, ferns, mosses, algae and lichen. These walls, some of which are over 100 years old, together with their green companions, provide unusual ecological and landscape values in the otherwise dull city environs (Jim, 1998c). The recent loss of high-grade members due to development pressures is lamentable, and highlights the sad lack of management and conservation of a rare natural-cum-cultural treasure of the community.

Overcoming fundamental constraints

Above-ground physical confinements

The inordinately tight and unfriendly town plan is a fundamental constraint to tree planting and it is diametrically inimical to a green-city notion. Associated with the basic straitjacket are the numerous above-ground impediments to crown expansion, such as narrow pavements, building awnings, unregimented traffic and advertisement signs. It is not uncommon to find post-planting intrusion into growing space, generating further conflict with existing trees (Jim, 1993a, 1997a–b). Passively waiting for plantable spaces to be eked out here and there can only bring insignificant, piecemeal and sluggish changes. Proactively finding, assigning and designating plantable spaces are needed to effect real and lasting improvements.

The solution will have to be realized in the long term by a gradual transformation of the old town plan. This has to be assiduously and consistently implemented through a setback of the building frontage from the roads for all redevelopment sites. A modest setback of about 3 m will be adequate to plant trees on a massive scale to permeate different quarters of the city and to completely transform the cityscape (Jim, 1999a). The roadside tree strip thus created will have to be guarded against above- and below-ground intrusions. Developers should be encouraged to comply by allowing the transfer of the plot ratio from the setback strip to the remainder of the site, thus no development potential whatsoever will be lost. Statutory measures are necessary to ensure that this bold plan will not be frustrated by intransigent parties.

Less-tangible institutional restrictions

The narrow roads in Hong Kong stifle tree planting in both the sense of the lack of physical space, as well as the need to satisfy the overwhelming road traffic requirements. There is an almost ubiquitous need to leave room for safety and sightline clearance to ensure traffic sign visibility, visibility between drivers and pedestrians, and unimpeded movement of people on the pavements. There are too many forbidden grounds, the full compliance of which would mean that few roadside sites could be legally planted. It will help if more flexibility could be exercised in resolving tree-traffic conflicts. There is a need to overhaul the mentality and practices in sharing the use of the cramped roadside space to serve a variety of key functions.

At present, many government and private-sector bodies at different levels are involved in a multivariate urban-tree system (Jim, 1987a), such as the formulation of greening policies, decision making, planting, management, protection and preservation. This situation may be a two-edged sword. There may be a multiplicity of ideas and initiatives, but there may also be a duplication of efforts and contradictions in purpose. The various forces should be joined together to ensure a commonality of objectives and actions. In this connection, a committee encompassing all players and stakeholders in urban trees could serve as a liaison and coordinating point to oversee the execution of a city-wide greening plan.

The general lack of community involvement in urban tree management in Hong Kong contrasts sharply with the case in other cities where such activities are routinely regarded as a joint venture between government and citizens. The expansion of publicity and awareness programmes, from the level of formal education to that of public education and community networking, will help to augment a sense of attachment and ownership of the urban trees. A concerned citizenry will demand more and better trees, and will provide necessary support to a greening programme.

Stifling underground utilities

The high-density city syndrome extends from the above-ground space to the underground environment (Jim, 1999b). The narrow roads carry vehicles and pedestrians above, and a substantial proportion of the city's infrastructure below. The buried utilities, in the form of pipes, cables, plus junction boxes and other cognate installations, occur profusely below the roads. Many lines run at a shallow depth directly below the pavement or at a short distance below the adjacent carriageway. Their omnipresence to a large extent has usurped the soil space for tree growth, thus making tree planting impossible along many road stretches (Jim, 1997a). They also restrict the root expansion of existing trees, and reduce the soil volume from which sustenance such as water and nutrients have to be derived (Jim, 1998d). A separation of the trees from the utility corridor, more easily implemented in new development areas, should reduce this conflict.

The underground services are frequently being excavated to insert new lines, repair old lines, divert old lines, disconnect old lines, and connect new lines. Locally, the trenching operations are particularly damaging to existing trees (Baines, 1994). This is because trenches tend to rigidly follow a straight line without any attempt to re-route when trees are encountered. Unlike in other places, in Hong Kong there is no guideline instructing the work crews on the ways to minimize excavation damage to trees. The relatively cheap, easy and mature technique of micro-tunnelling, widely practised in many cities to avoid digging near trees, has not been introduced in Hong Kong (National Joint Utilities Group, 1995; Harris et al., 1999). Many roadside trees have to endure repeated root destruction due to trenching, and hence they perform poorly and will decline and die prematurely. The loss of the amenity functions of the affected trees, the need for more tree maintenance, and their early demise, may be translated into significant monetary losses to the community. The cost of using a non-digging technique could easily be recouped by having better and longer-living trees that require less care. This sort of green accounting vividly illustrates that attempts to cut corners by utility companies will result in more costly operations, but the extra burdens will be shouldered by the community.

Poor urban soil quality

The subsurface portion of urban trees is often neglected and taken for granted (Watson and Neely, 1994; Neely and Watson, 1998). Soils at urban planting sites in Hong Kong are highly disturbed and contaminated almost everywhere, more so than in other cities (Bullock and Gregory, 1991; Craul, 1992). Very few sites have their original soil and vegetation cover left intact. Natural soil profiles have been quite thoroughly destroyed. Most land surfaces have been either truncated or filled, and invariably they have received foreign earth materials, which are mainly derived from construction rubble mixed with all sorts of artifacts. The hill terraces were constructed by cut and fill to form level platforms on which roads or buildings were established. The reclaimed lands were derived from filling either with demolition wastes or from earth borrowed from land- or marine-based borrow areas. The composition and properties of the resulting soils are unfavourable to plant growth. Recent research findings on local urban soils confirm that they are usually of very poor quality in physical, chemical and biological terms (Jim, 1998d–h, 1999b). Yet the landscape planting practices in Hong Kong often make use of the existing "site" soil to grow amenity vegetation with little or no amelioration.

The urban soils are beset by a number of common problems which limit plant growth in different ways (Jim, 1993c). They usually contain an excessive amount of stone and sand which is inimical to the storage of available moisture, and to the supply and storage of available nutrients. The coarse fragments also act as a physical hindrance to root growth. The soils are often heavily compacted during the construction phase, or deliberately compressed to meet load-bearing engineering requirements in road construction practices, or subject to human-foot trampling pressures (Jim, 1993d). The soil surfaces may be sealed and the structure degraded, resulting in a lack of porosity for aeration, infiltration, drainage, and moisture storage. At many planting sites, the soil depth (the solum) is often too shallow for normal root-system expansion.

The nutrient stock of urban soils is usually inadequate in terms of the total quantity as well as the ability to hold nutrients in readily available reserve forms. Commonly, the soils are contaminated with calcareous materials derived from cement, concrete, plaster and lime associated with construction debris. Consequently, the soil reaction is often alkaline, with the pH going up to 9 to 10 in extreme cases which may induce nutrient imbalance problems (Messenger, 1986). Most humid-tropical plants have adapted to the acidic soil environment akin to the normal pH range (of around 5) of natural soils in Hong Kong. Some soils at roadsides or specific sites affected by industrial activities may be polluted by heavy metals and toxic organic compounds.

To have a high-quality greening programme, it is necessary to look after both the above- and the below-ground environment for plant growth. Present practices tend to neglect the substrate, which is believed to be the major determinant of poor tree performance in Hong Kong. A tree's soil requirement is not that difficult to satisfy — it needs about 1 m deep layer of good soil, with as much lateral extent as possible, for healthy growth (Perry, 1994). In considering the provision of an adequate volume of soil, it should be noted that a tree normally spreads its roots to a distance of at least twice as wide as its crown. Thus a tree with a 10 m diameter crown will extend its roots to an area 20 m wide. Measures should be taken to ensure that planting sites as far as practicable will satisfy this minimum soil requirement. At the linear roadside sites, a continuous corridor of good soil should be provided down to 1 m depth to ensure that the roots at least can spread out in two directions (along the road lineament). If the site soil is too poor to be improved by amendments, it should be replaced with a good-quality soil mix prepared according to clearly-defined specifications.

Soil replacement should be more liberally adopted in view of the common occurrence of poor site soils.

Urban tree governance

The present urban-tree management system, with responsibilities scattered amongst a number of government departments and with little coordination of private-sector initiatives, can be improved to bring better trees in more places. The Leisure and Cultural Services Department (LCSD) supervises tree management in most urban areas, including roadsides, public parks, gardens and other amenity venues, and government building grounds. It has a team of amenity officers, some of whom have been assigned to look after horticultural duties. In addition, other government departments are involved in different aspects of urban tree management. They include: (1) The Planning Department which scrutinizes development applications which may affect or plant trees, and can impose tree preservation and planting as approval conditions; (2) Agriculture, Fisheries and Conservation which plants and cares for trees in the countryside, and is consulted on trees affected by development projects that occupy undeveloped lands; (3) The Housing Department which plants and cares for trees in public housing estates that accommodate about half of the total population; (4) The Highways Department which builds roads and in the process often plants trees at roadsides to be handed over to the LCSD for long-term maintenance, but may sometimes be also involved in massive tree felling; (5) The Lands Department which controls all land administration matters and is consulted on the protection of existing trees situated at sites that are earmarked for development; (6) The Territory Development Department which oversees the development of new towns is involved in tree planting and preservation, to be handed over to the appropriate authorities for long-term care; (7) The Home Affairs Department in conjunction with the District Councils to carry out district-level and small-scale environmental improvement works that often involve tree planting; (8) The Architectural Services Department which designs and implements government development projects that sometimes include landscape planting.

There is obviously a surfeit of tree agencies in Hong Kong, each participating in a given segment of the overall urban-tree management system. The LCSD is the major player as its responsibility includes the largest number of trees in the built-up areas. The Housing Department is another key player as all lands and their constituent greeneries lying within public housing estates fall within its sphere of responsibility. Other departments each makes contributions, ranging from proffering expertise opinions and comments to actual planting and care. The private developers who could play a pivotal role in city greening, meanwhile, remain untapped and uncoordinated. The multiple management authorities could bring a diversity of initiatives which could theoretically enrich the tree programme. The cross-fertilization of ideas and synergistic benefits, however, cannot materialize if there is little coordination and cooperation between the disparate parties. The demarcation of duties and responsibilities amongst the plethora of stakeholders is rather nebulous. There is no clear and unambiguous official authority playing a leadership role on tree matters within the administration. To cultivate a culture of collaboration and mutual support, it may be necessary to set up a body similar to the *tree council* adopted in other countries. A city-wide *master greening plan* should be instituted as soon as possible (Jim, 1999a) to provide the overriding working framework, to furnish clear objectives and directions, and to knit together the expertise, experience and efforts from different quarters both within and without the government. Mobilizing wide community support and participation will be a necessary ingredient for a successful greening programme.

There is a need to upgrade the standard of tree care to catch up with that achieved in other countries (Mattheck and Breloer, 1994; Mattheck and Kubler, 1995; Gilman, 1997; Kozlowski and Pallardy, 1997; Lloyd, 1997; Watson and Himelick, 1997). A summary of the main management issues, problems and inadequacies, and recommended actions, is given in Table 1. The acquisition of relevant knowledge and the application of modern concepts and practices will enhance tree care. The improvement package will have to be encompassing, covering the nurturing or acquisition of planting materials, the selection of species to enhance the cityscape and to match different site conditions (Jim, 1990b–d), and timely and preemptive tree maintenance (Shigo, 1991; Harris et al., 1999) with a view to living with the periodical onslaught of natural hazards, especially typhoons (Jim and Liu, 1997). It would help if tree care could adopt a cradle-to-grave mentality so that both newly planted as well as mature and over-mature trees are given adequate attention. The massive and avoidable damages to trees due to construction activities (Jim, 1988; National House Builders Corporation, 1992) and ubiquitous trenching should be minimized by more

Table 1. A synopsis of the major urban-tree management issues in Hong Kong with reference to specific problems or inadequacies and recommended actions

	Management issue	Problem or inadequacy	Recommended action
1	Plantable space and site quality [One cannot make omelettes without breaking eggs]	Inordinately tight town plan	Open up town plan by instituting city-wide building setback
		Cramped planting sites with many constraints	Designate dedicated tree strip
		– aboveground physical and environmental limitations	Control intrusion into tree growing space
		– belowground profusion of underground utilities	Regiment underground utilities in separate zone
		– extremely poor quality soil	Design roadside soil corridor
		– inadequate soil volume for root expansion	Replace poor site soil with good quality soil mix
			Maintain site quality
2	Multiple tree management authorities [Too many cooks spoil the broth]	Nine government departments involved in various phases of tree management	Designate one department as official tree authority and leader on tree matters
		Nebulous demarcation of duties and responsibilities	Assign duties and responsibilities unambiguously to the participating departments
		No direction or encouragement to private sector greening initiatives	Reduce the number of departments involved in tree matters
		Sometimes conflicting aims and objectives	Provide guidance and incentives to private sector greening
		Duplication of efforts and wasteful of resources	Establish a coordinating body to oversee city-wide greening initiatives and activities
3	City-wide master greening plan [See the trees without seeing the forest]	No synoptic and comprehensive greening plan for the whole city	Promulgate a city-wide master greening plan
		Greening endeavours often conducted at district level	Appoint a high-level body to oversee implementation of the greening plan
		Lack of direction and purpose regarding greening for both government and private-sector projects	Give encouragement and incentives to private sector to help realize the greening plan
4	Dedicated tree ordinance [Take the law into the trees' hands]	No dedicated tree ordinance	Enact a dedicated tree ordinance
		Reliance on a collection of general statues and administrative measures	Designate a clear authority to enforce the tree ordinance
		Ineffectual statutory protection of trees	Set penalties on advertent tree damage to a realistic level
		Legal vacuum for tree protection at some sites	Regulate tree planting requirements for different land uses and development projects

Table 1. (Cont'd)

Management issue	Problem or inadequacy	Recommended action
5 Quality planting material *[Better nip it in the bud]*	Poor quality nursery practice and products – lack of vigour, sparse foliage – crossed branches, v-crotch – unbalanced crown – crooked and curved trunk – multiple stems, wounded or decayed stem Long-term liabilities and potential hazards	Select planting materials carefully Screen rigorously at the nursery stage Adopt stringent specifications and standards Inspect systematically and thoroughly Reject poor-quality planting materials Replace existing poor trees as soon as possible
6 Innovative species selection *[Variety is the spice of life]*	Excessive domination by – exotic species – bland foliage and evergreen species – species with small final dimensions Too many palms Adherence to familiar cohort – too focused on common species Mismatch between species and site conditions	Enhance diversity of species and cultivars – attractive flowers – seasonal changes – different tree forms – large final dimensions Ensure better match between species and site conditions Adopt more outstanding native and exotic ornamentals Set up plant introduction unit to screen suitable candidates
7 Timely treatment *[Procrastination is the thief of time]*	Lack of systematic and regular tree inspection Problems allowed to linger and deteriorate Response rather than proactive approach to tree care	Inspect regularly by qualified arborist Identify problems in good time Prioritize tree-care actions Pay prompt attention to exigencies Win confidence and trust of citizens
8 Preemptive maintenance *[A stitch in time saves nine]*	Nature purges trees – at wrong time and place – weak trees are more affected	Reduce nature's rigorous but haphazard screening work Adopt prognosis practice and timely maintenance
9 Living with natural hazards *[God's mill grinds slow but sure]*	Occasional typhoon onslaughts – natural selective force – periodic purging of the weaklings Neglected preventive tree care Messy and hazardous consequences	Apply timely care and maintenance Identify the weakling and take preventive measures Set up a hazard-tree assessment programme Select wind-tolerant species at exposed sites Nurture strong trees which are more tolerant of strong winds

	Problems	Recommendations
10 Long-term enterprise [From cradle to grave]	Attention concentrated on initial establishment period Inadequate care for mature trees Senile and battered mature trees often neglected	Go beyond the initial establishment period – attention to mature trees – especially champion specimens Institute regular tree inspection Introduce programmed tree removal Provide guidance and enforce good arboricultural practices
11 Containing construction and trenching damages [More in sorrow than in anger]	Cursory attitude towards trees Inadequate knowledge about tree protection Inadequate tree survey and site evaluation Poor workmanship and inadequate supervision Building plans disregard existing trees Unaware of modern tree-friendly techniques Distorted property value belittles tree value Misunderstanding or ignoring tree protection requirements	Require high-quality tree survey report Demand sympathetic building plan to preserve trees Upgrade professional standard tree protection and preservation Stipulate the use of micro-tunnelling (trenchless) machines for all excavations near large and valuable trees Statutory protection of trees Give realistic monetary value to trees Adopt proactive programme of protection
12 Preserving champion specimens [Big oaks from little acorns grow]	Lack of legal status and protection – merely 360 in study area Continued damage and felling Poor recognition of their heritage value	Introduce statutory protection as living heritage Raise community awareness and recognition Provide high-level arboricultural care
13 Information accumulation and exchanges [Never shall the twain meet?]	Tree knowledge and practices derived from western and temperate-latitudes Shortage of specific scientific information on tropical tree species and cultivars Sluggish transfer of knowledge Lack of adaptation of knowledge for tropical	Develop tropical urban forestry Initiate appropriate research Expand knowledge repertoire Network scientists and practitioners Form tropical amenity tree league Initiate regular cooperation and exchanges arboriculture

Table 1 (Cont'd)

Management issue	Problem or inadequacy	Recommended action
14 Suggested research directions *[Longing for the moment of truth]*	Insufficient research on local and tropical trees	Conduct pure and applied research on:
	Little research on tree planting and care methods for local environment and trees	– environmental benefits and functions
	Inadequate funding and support for research	– nursery production
	Lack of cooperative research with local universities	– evaluation of native tropical species
	Importance of research not appreciated	– hazardous trees
	No critical mass for tree research	– urban soil for tree growth
		– planting material standards
		– protection of champion trees
		– tree ordinance
		– institutional setup to provide tree sites
		– community involvement and participation
15 Adopting arboricultural advances *[Necessity is the mother of invention]*	Recent advances in arboricultural concepts and practices not earnestly adopted	Apply new tree concept to tree care
	Modern equipment in tree care not promptly evaluated for local application	– compartmentalization of decay in trees (CODIT)
		– body language and tree hazards
		– urban soil science for tree growth
		Test and adopt modern equipment to aid tree work
16 Personnel training *[The greatest gift of life is education]*	Lack of training to a high level	Strengthen overseas and local training
	Lack of continued professional training	Train up to degree and postgraduate level
	Lack of a clear professional leader in the tree team	Improve on-the-job training
	Detachment from local and overseas professionals and academics	Establish dedicated tropical urban forestry training
		Cooperate with tropical countries
		Encourage continued professional development

stringent standards in tree protection and trenchless micro-tunnelling techniques. The quality of transplanting large trees should be upgraded to international levels (Jim, 1995). The small cohort of champion trees, rare gems and the natural heritage of the city requires special protection and care (Randall and Clepper, 1977; Jim, 1994h-i, 1998i).

The present legal protection of trees, including the top-echelon specimens, is grossly ineffectual. To a large extent this is due to the absence of a tree ordinance and hence the need to rely on a collection of general statues and administrative measures to regulate human behaviour towards trees. Such a state of affairs is out of tune with the desire to upgrade greening and to protect the existing stock (Profous and Loeb, 1990). The enactment of a dedicated tree ordinance is strongly urged in order to place this particular realm of tree management on a higher plane (Merullo and Valentine, 1992). The recent advances in arboricultural knowledge and practice could be more earnestly introduced into Hong Kong and tested for their applicability to local tress and the cramped urban environment. An assessment of modern tree-care equipment for suitability to local tree care could be more earnestly pursued. Research could be more generously supported in both financial and institutional terms to build up a knowledge base on tropical amenity trees and tree-care methodology. A network of researchers and practitioners could help to communicate and share ideas and know-how. The tree staff could be more highly educated and be given regular continued professional training to keep up with the latest advances.

Conclusion

That many aspects of our infrastructure are of first-world standards, yet our green infrastructure remains in the third-world bracket, has to be more emphatically stressed and broadly recognized. There is a desire to augment the quantity and quality of amenity vegetation in urban Hong Kong, as part and parcel of recent attempts to build a livable, sustainable and international city (Lennard and Lennard, 1987). The realization of this bold plan cries out for changes in the existing management and governance structure, without which it is anticipated that the greening efforts will continue to be piecemeal, limited in scope, and insignificant in upgrading the cityscape. The myth that a high-density city such as Hong Kong cannot possibly harbour much greenery should be dispelled once and for all. The present status of the urban forest indicates some constraints and pitfalls which need to be rectified in order to forge ahead with a more ambitious greening programme. Staying in this rut will not take us anywhere other than trapping us in the quagmire of mediocrity. We need a clean departure from the district-local tunnel-vision mode of greening to a territory-wide, long-term and green-metropolis vision. A macro-scale master green plan, directed at the top levels of the administration, is a crucial precursor if not a precondition for a successful endeavour.

A prime consideration has to be the lack of quality planting sites in the unduly cramped town plan which must be opened up in order that high-calibre greenery may be more liberally inserted into the city. Once the planting sites are secured, we have to fill them with high-calibre planting materials that are appropriate to the site conditions, and provide them with high-calibre tree care through their trying tenure. Meanwhile, we have to enhance our ability to protect existing trees against construction, trenching, typhoons and other forces of destruction. The top quality champion trees in particular deserve special attention so as to guard them against insidious and overt incursions. Effective tree management requires quality personnel equipped with quality education, support and equipment. All these efforts require sustained resources, uncompromising professionalism, strong leadership, and above all an unswerving determination and an unrelenting dedication to the worthy cause. Overall, greening a city, especially an existing and old city, demands a huge amount of efforts that must be meticulously coordinated without respite at all levels. If we want to turn the pipe dream into reality, we have to be prepared to put greening at the top of the community's agenda, to go the whole hog, to adopt a long-term perspective, and to persevere along a long and bumpy road to the elusive utopia.

References

Agriculture and Fisheries Department. "Check List of Hong Kong Plants." *Agriculture and Fisheries Department Bulletin 1 (Revised)*. Hong Kong Herbarium, 1993, 159 pp.

Baines, C. "Trenching and Street Trees." *Arboricultural Journal,* 18 (1994), pp. 231–236.

Bradshaw, A. D, Hunt, B. and Walmsley, T. *Trees in the Urban Landscape: Principles and Practice*. London: Spon, 1995, 272 pp.

Bullock, P. and Gregory, P. J. (eds.). *Soils in the Urban Environment*. Oxford: Blackwell, 1991, 174 pp.

Census and Statistics Department. *Annual Digest of Statistics 1999*. Hong Kong: Hong Kong Government, 2000, 291 pp.

Cole, L. "Urban Opportunities for a More Natural Approach." In *Ecology and Design in Landscape*, edited by A. D. Bradshaw, D. A. Goode and E.H.P. Thorp, pp. 417–431. Oxford: Blackwell, 1986.

Corlett, R. T. "Environmental Forestry in Hong Kong: 1871–1997." *Forest Ecology and Management,* 116 (1999), pp. 93–105.

Craul, P. J. *Urban Soil in Landscape Design*. New York: John Wiley and Sons, 1992, 396 pp.

Flink, C.A. and Searns, R.M. *Greenways: A Guide to Planning, Design, and Development*. Washington DC: Island Press, 1993, 351 pp.

Gilman, E.F. *An Illustrated Guide to Pruning*. Albany, NY: Delmar, 1997, 78 pp.

Grey, G.W. *The Urban Forest: Comprehensive Management*. New York: John Wiley and Sons, 1996, 156 pp.

Grey, G.W. and Deneke, F.J. *Urban Forestry*, 2nd ed. New York: John Wiley and Sons, 1986, 299 pp.

Harris, R.W., Clark, J.R. and Matheny, N.P. *Arboriculture: Integrated Management of Landscape Trees, Shrubs and Vines*, 3rd ed. New Jersey: Prentice Hall, 1999, 687 pp.

Howard, E. *Garden Cities of Tomorrow*. London: Faber and Faber, 1902, 168 pp.

Hong Kong Museum of History. *The Hong Kong Album: A Selection of the Museum Historical Photographs*. Hong Kong: Urban Council, 1982, 99 pp.

Jim, C.Y. "The Country Parks Programme and Countryside Conservation in Hong Kong." *The Environmentalist,* 6 (1986), pp. 259–270.

Jim, C.Y. "The Status and Prospects of Urban Trees in Hong Kong." *Landscape and Urban Planning,* 14 (1987a), pp. 1–20.

Jim, C.Y. "Land Use and Amenity Trees in Urban Hong Kong." *Land Use Policy,* 4 (1987b), pp. 281–293.

Jim, C.Y. "Preservation of a Large Chinese Banyan on a Construction Site." *Journal of Arboriculture,* 14 (1988), pp. 176–180.

Jim, C.Y. "The Distribution and Configuration of Tree Cover in Urban Hong Kong." *GeoJournal,* 18 (1989a), pp. 175–188.

Jim, C.Y. "Tree Canopy Cover, Land Use and Planning Implications in Urban Hong Kong." *Geoforum,* 20 (1989b), pp. 57–68.

Jim, C.Y. "Tree Canopy Characteristics and Urban Development in Hong Kong." *Geographical Review,* 79 (1990a), pp. 210–225.

Jim, C.Y. *Trees in Hong Kong: Species for Landscape Planting*. Hong Kong: Hong Kong University Press, 1990b, 434 pp.

Jim, C.Y. "Selection of Tree Species for Urban Plantings in Tropical Cities." In *Proceedings 19th World Congress of the International Union of Forest Research Organizations*, pp. 236–247. Montreal: Division 1, Volume 1, 1990c.

Jim, C.Y. "Evaluation of Tree Species for Amenity Planting in Hong Kong." *Arboricultural Journal,* 14 (1990d), pp. 27–44.

Jim, C.Y. "Diversity of Amenity Species in Hong Kong." *Quarterly Journal of Forestry,* 55 (1991), pp. 233–243.

Jim, C.Y. "Provenance of Amenity-tree Species in Hong Kong." *Arboricultural Journal,* 16 (1992a), pp. 11–23.

Jim, C.Y. "Tree-habitat Relationships in Urban Hong Kong." *Environmental Conservation,* 19 (1992b), pp. 209–218.

Jim, C.Y. "Trees and High-density Urban Development: Opportunities Out of Constraints." *Habitat International,* 17 (1993a), pp. 1–17.

Jim, C.Y. "Trees and Landscape of a Suburban Residential Neighbourhood in Hong Kong." *Landscape and Urban Planning,* 23 (1993b), pp. 119–143.

Jim, C.Y. "Massive Tree-planting Failures Due to Multiple Soil Problems." *Arboricultural Journal,* 17 (1993c), pp. 309–331.

Jim, C.Y. "Soil Compaction as a Constraint to Tree Growth in Tropical and Sub-tropical Urban Habitats." *Environmental Conservation,* 20 (1993d), pp. 35–49.

Jim, C.Y. "Urban Renewal and Environmental Planning in Hong Kong." *The Environmentalist,* 14 (1994a), pp. 163–181.

Jim, C.Y. *Urban Tree Survey 1994: Roadside Trees Managed by the Urban Council: Executive Summary.* Hong Kong: Urban Council, 1994b, 37 pp.

Jim, C.Y. *Urban Tree Survey 1994: Roadside Trees Managed by the Urban Council: Survey Results and Recommendations.* Hong Kong: Urban Council, 1994c, 470 pp.

Jim, C.Y. *Urban Tree Survey 1994: Roadside Trees Managed by the Urban Council: Summary Tables for the Tree Census.* Hong Kong: Urban Council, 1994d, 243 pp.

Jim, C.Y. *Urban Tree Survey 1994: Roadside Trees Managed by the Urban Council: Survey Data for the Tree Census.* Hong Kong: Urban Council, 1994e, 197 pp.

Jim, C.Y. *Urban Tree Survey 1994: Roadside Trees Managed by the Urban Council: Summary Tables and Survey Data for Potential Planting Sites.* Hong Kong: Urban Council, 1994f , 197 pp.

Jim, C.Y. *Urban Tree Survey 1994: Roadside Trees Managed by the Urban Council: Map Sheets (1:1000) Showing the Locations of Existing Roadside Trees and Potential Planting Sites.* Hong Kong: Urban Council, 1994g, 169 pp.

Jim, C.Y. *Champion Trees in Urban Hong Kong.* Hong Kong Flora and Fauna Series. Hong Kong: Urban Council, 1994h, 294 pp.

Jim, C.Y. "Evaluation and Preservation of Champion Trees in Urban Hong Kong." *Arboricultural Journal,* 18 (1994i), pp. 25–51.

Jim, C.Y. "Transplanting Two Champion Specimens of Mature Chinese Banyans." *Journal of Arboriculture,* 21 (1995), pp. 289–295.

Jim, C.Y. *The Multiple Values of Urban Trees.* Hong Kong: Urban Council, 1996a, 79 pp.

Jim, C.Y. "Roadside Trees in Urban Hong Kong: Part I Census Methodology." *Arboricultural Journal,* 20 (1996b), pp. 221–237.

Jim, C.Y. "Roadside Trees in Urban Hong Kong: Part II Species Composition." *Arboricultural Journal,* 20 (1996c), pp. 279–298.

Jim, C.Y. "Roadside Trees in Urban Hong Kong: Part III Tree Size and Growth Space." *Arboricultural Journal,* 21 (1997a), pp. 73–88.

Jim, C.Y. "Roadside Trees in Urban Hong Kong: Part IV Tree Growth and Environmental Condition." *Arboricultural Journal,* 21 (1997b), pp. 89–99.

Jim, C.Y. "Impacts of Intensive Urbanization on Trees in Hong Kong." *Environmental Conservation,* 25 (1998a), pp. 146–159.

Jim, C.Y. "Pressure on Urban Trees in Hong Kong: Pervasive Problems and Possible Amelioration." *Arboricultural Journal,* 22 (1998b), pp. 37–60.

Jim, C.Y. "Old Stone Walls as an Ecological Habitat for Urban Trees in Hong Kong." *Landscape and Urban Planning,* 42 (1998c), pp. 29–43.

Jim, C.Y. "Soil Characteristics and Management in an Urban Park in Hong Kong." *Environmental Management,* 22 (1998d), pp. 683–695.

Jim, C.Y. "Urban Soil Characteristics and Limitations for Landscape Planting in Hong Kong." *Landscape and Urban Planning,* 40 (1998e), pp. 235–249.

Jim, C.Y. "Nature and Management of Urban Soils in Hong Kong." *Transactions 16th International Congress of Soil Science,* Section on Urban Soil, International Society of Soil Science, Montpellier, France, 1998f, pp. 1–12.

Jim, C.Y. "Soil Compaction at Tree Planting Sites in Urban Hong Kong." In *The Landscape Below Ground II,* edited by G.W. Watson and D. Neely, pp. 166–178. Savoy, IL: International Society of Arboriculture, 1998g.

Jim, C.Y. "Physical and Chemical Properties of a Hong Kong Roadside Soils in Relation to Urban Tree Growth." *Urban Ecosystems,* 2 (1998h), pp. 171–181.

Jim, C.Y. *Champion Trees in Urban Hong Kong,* rev. ed. Hong Kong Flora and Fauna Series. Hong Kong: Urban Council, 1998i, 342 pp.

Jim, C.Y. "A Planning Strategy to Augment the Diversity and Biomass of Roadside Trees in Urban Hong Kong." *Landscape and Urban Planning,* 44 (1999a), pp. 13–32.

Jim, C.Y. "Trees and the Greening of Hong Kong's Urban Landscape: Subaerial and soil constraints." In *Remediation and Management of Degraded Lands,* edited by M.H. Wong, J.W.C. Wong and A.J.M. Baker, pp. 209–224. Boca Raton, FL: Lewis Publishers/CRC Press, 1999b.

Jim, C.Y. *Trees in Major Urban Parks in Hong Kong: Census Results and Management Recommendations.* Hong Kong: Urban Council, 2000a, 193 pp.

Jim, C.Y. *Trees in Major Urban Parks in Hong Kong: Summary Tables and*

Figures for Parks on Hong Kong Island. Hong Kong: Urban Council, 2000b, 405 pp.

Jim, C.Y. *Trees in Major Urban Parks in Hong Kong: Summary Tables and Figures for Parks in Kowloon.* Hong Kong: Urban Council, 2000c, 319 pp.

Jim, C.Y. *Trees in Major Urban Parks in Hong Kong: Tree Location Maps and Species Indices.* Hong Kong: Urban Council, 2000d, 432 pp.

Jim, C.Y. and Liu, H.H.T. "Storm Damage on Urban Trees in Guangzhou, China." *Landscape and Urban Planning,* 38 (1997), pp. 45–59.

Kelcey, J.G. "The Green Environment of Inner Urban Areas." *Environmental Conservation,* 5 (1978), pp. 197–203.

Kozlowski, T.T. and Pallardy, S.G. *Physiology of Woody Plants,* 2nd ed. New York: Academic Press, 1997, 411 pp.

Lawrence, H.W. "Origins of the Tree-lined Boulevard." *The Geographical Review,* 78 (1998), pp. 355–374.

Lennard, S.H.C. and Lennard, H.L. *Livable Cities. People and Places: Social and Design Principles for the Future of the City.* Southampton, NY: Gondolier Press, 1987, 166 pp.

Lloyd, J. (coordinating author). *Plant Health Care for Woody Ornamentals.* Savoy, IL; International Society of Arboriculture, 1997, 223 pp.

Mattheck, C. and Breloer, H. *The Body Language of Trees: A Handbook for Failure Analysis.* Research for Amenity Trees No. 4, edited by D. Lonsdale, translated from the German by R. Strouts. London: HMSO, 1994, 240 pp.

Mattheck, C. and Kubler, H. *Wood — the Internal Optimization of Trees.* Berlin: Springer, 1995, 129 pp.

Merullo, V.D. and Valentine, M.J. *Arboriculture and the Law.* Savoy, IL: International Society of Arboriculture, 1992, 110 pp.

Messenger, S. "Alkaline Runoff, Soil PH and White Oak Manganese Deficiency." *Tree Physiology,* 2 (1986), pp. 317–325.

Miller, R.W. *Urban Forestry: Planning and Managing Urban Greenspaces,* 2nd ed. Englewood Cliffs, NJ: Prentice Hall, 1997, 502 pp.

Mumford, Lewis. *The City in History: Its Origins, its Transformations, and its Prospects.* London: Secker and Warburg, 1961, 657 pp.

National Joint Utilities Group. *Guidelines for Planning, Installation and Maintenance of Utility Services in Proximity to Trees.* Publication Number 10. London: NJUG, 1995, 23 pp.

National House Builders Corporation. *NHBC Standards: Building Near Trees.* Amersham, Bucks, UK: NHBC, 1992, various paging.

Neely, D. and Watson, G.W. (eds.). *The Landscape Below Ground II.* Proceedings of an International Workshop on Tree Root Development in Urban Soils, San Francisco, March 5 and 6, 1998. Savoy, IL: International Society of Arboriculture, 1998, 265 pp.

Nicholson-Lord, D. *The Greening of the Cities*. London: Routledge and Kegan Paul, 1987, 270 pp.

Platt, R.H., Rowntree, R.A. and Muick, P.C. *The Ecological City: Preserving and Restoring Urban Biodiversity*. Amherst, MA: University of Massachusetts Press, 1994, 291 pp.

Perry, T.O. "Size, Management and Design of Tree Planting Sites." In *The Landscape Below Ground*, edited by G.W. Watson and D. Neely, pp. 3–15. Savoy, IL: International Society of Arboriculture, 1994.

Profous, G.V. and Loeb, R.E. "The Legal Protection of Urban Trees: A Comparative World Survey." *Journal of Environmental Law*, 2 (1990), pp. 179–193.

Randall, C.E. and Clepper, H. *Famous and Historic Trees*. Washington, DC: American Forestry Association, 1977, 90 pp.

Schmid, J.A. *Urban Vegetation: A Review and Chicago Case Study*. Research Paper Number 161. Chicago, IL: Department of Geography, University of Chicago, 1975, 266 pp.

Shigo, A.L. *Modern Arboriculture: A Systems Approach to the Care of Trees and their Associates*. Durham, NH: Shigo and Trees Associates, 1991, 424 pp.

Watson, G.W. and Himelick, E.B. *Principles and Practice of Planting Trees and Shrubs*. Savoy, IL: International Society of Arboriculture, 1997, 199 pp.

Watson, G.W. and Neely, D. (eds.). *The Landscape Below Ground*. Proceedings of an International Workshop on Tree Root Development in Urban Soils, Morton Arboretum, Chicago, September 30 to October 1, 1993. Savoy, IL: International Society of Arboriculture, 1994, 222 pp.

Zhuang, X., Xing, F. and Corlett, R.T. "The Tree Flora of Hong Kong: Distribution and Conservation Status." *Memoirs of the Hong Kong Natural History Society*, 21 (1997), pp. 69–126.

Meeting Higher Order Needs in Municipal Solid Waste Management: A Comparison between Hong Kong and Guangzhou

Shan-shan Chung and Carlos Wing-hung Lo
The Hong Kong Polytechnic University

Introduction

Just as human beings have different levels of needs, there are different levels of waste management demands in a society. Figure 1 depicts the widely adopted waste management hierarchy. In the context of sustainable waste management, the order of preference runs from the top to the bottom, i.e., waste avoidance is preferred to recycling and so on (United Nations Conference on Environment and Development, 1992). Throughout the life cycle of a waste substance various measures can be used to reduce its environmental impact and these can be grouped under three generic approaches: prevention, recovery and pollution control. In this paper, the term "higher order waste management needs/measures" is used to refer to the more preferred waste management options, such as avoidance, prevention and recovery. Higher order waste management measures are able to achieve resource conservation, energy conservation and prevention of the destruction of the ecological environment due to the occupation of land for waste treatment/disposal and related pollution discharge. The end of pipe pollution control measures, such as safe disposal, waste collection and related processes, are referred to as "lower order waste management measures".

One key insight to attain environmental sustainability as brought up by *Agenda 21* is that of education — to provide all citizens with the

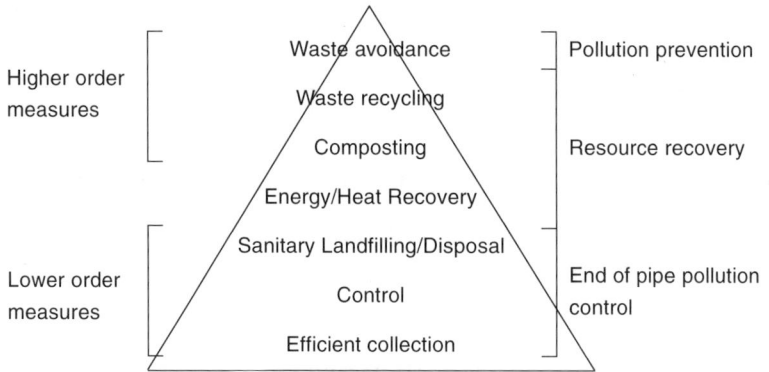

Figure 1. The waste management hierarchy and levels of waste
management needs

information and education they need to fully participate in the transition to a more sustainable lifestyle. Environmental citizenship is the state of education for sustainability (Hawthorne and Alabaster, 1999). Sustainable waste management similarly requires public participation, especially in higher order waste management.

Despite general support to the waste management hierarchy, empirical evidence indicates that few countries are able to follow the hierarchy closely. While some may already have the lower order needs taken care of, the more preferred higher order measures are neglected. For instance, in Greece, despite almost complete coverage in regular waste collection services, municipal recycling programs are the exception rather than the norm (Agapitidis and Frantzis, 1998). In other developed countries, although higher order waste management measures were instituted, they were however dealt with only *after* lower order needs were satisfied. For example, in Germany, the Waste Disposal Act that required the provision of pollution control measures in waste facilities was implemented in 1972 but the first law to institutionalize waste avoidance and recycling measures was not promulgated until 1986 (*Act on the Avoidance and Management of Waste*) (Szelinski, 1988) — fourteen years after the Waste Disposal Act.

In developing areas, the implementation of the waste management hierarchy is even more disappointing. Neither lower order or higher order needs are adequately met. Blight and Mbande (1996) have painted a picture of ineffective waste collection coupled with informal waste scavenging but a lack of formal waste recovery measures in developing countries. Danteravanich and Siriwong (1998) describe waste management in southern Thailand as ineffective in controlling pollution from waste disposal and barely including any formal recycling of solid waste. The situation is worse in Tanzania where only 30% of the waste generated is collected and disposed of. Moreover, disposal sites in Tanzania are no more than open dumps. Actions to improve collection and disposal and to institute recycling have yet to be found (Yhdego, 1995:2). Similarly, in Lima, the capital of Peru, none of the waste management alternatives on the hierarchy has been carried out satisfactorily. Only about 60% of the waste generated in Lima is collected and only 30% of the waste collected will reach a disposal site (Diaz, 1999). In the capital of Vietnam, local waste management services are constrained by a lack of resources and only about two-thirds of the waste is collected. Waste recovered for recycling is mainly done by scavengers and not in any planned manner (Urban Environment Company, 1995).

Nevertheless, there are examples of higher order measures being instituted in developing areas. For instance, in the case of Lima, recycling programs were put in place, though not until the US Agency for International Development piloted a solid waste management project in the 1990s (Diaz, 1999). In Hanoi, a pilot composting facility project, including a pre-composting process to sort and recover metals and plastics, was financed by the United Nations Development Program (Urban Environment Company, 1995). It should be noted that in both cases the impetus to regularize these higher order measures has been external.

In sum, a prioritization conflict between government empirical behaviour and the waste management hierarchy is noted in both developed and less developed areas. Governments tend to regard collection and safe disposal as more fundamental and the priority in resource allocation is given to lower order needs rather than to higher order needs.

Is this conflict also found in China, the most populated country in the world? What necessitates that after lower order needs are satisfied, higher order measures will follow? And how quickly can more sustainable waste management be put forward after lower order needs are met? Most important of all, if the waste management hierarchy is considered a sensible and tenable theoretical framework (Schall, 1992), how can such conflicts be resolved?

This paper aims to answer the above questions through a comparative evaluation of two municipal solid waste (MSW) management systems, Hong Kong Special Administrative Region (HK) and Guangzhou (GZ). There are four parts in this paper. The first part provides the socio-economic characteristics and background information on waste management in the two cities. Part two is an analysis of five aspects of waste management: policy ideology, policy content, regulatory process, public participation and policy consequences. Part three is an evaluation of the extent to which various waste management needs in the two systems are met. The following five criteria will be used in the evaluation: scope of waste managed, environmental impact, implementation effectiveness, cost-effectiveness and use of a participatory approach. Part four is a concluding section that attempts to answer the questions set out in the previous paragraph concerning how and why there are conflicts. This leads to a discussion on the constraints to meeting higher and lower order waste management needs in the two cities.

Background

This part is composed of a number of introductory sections on the socio-economic, institutional and statistical aspects of the waste management systems of GZ and HK.

Socio-economic snapshot

Table 1 shows some important demographic and socio-economic data of the two cities. In short, HK has a higher population density and is more affluent than Guangzhou.

MSW generation

Waste matters can be classified according to the sources of their generation (domestic, commercial, industrial, clinical, etc.). In GZ, MSW is referred

Table 1. Demographic and socio-economic data of HK and GZ

	Guangzhou	Hong Kong
Population in 1998	3.99 million	6.81 million
Area (in km^2)	1443.6	1092
GDP per capita in 1998	¥32,514 RMB (HK$30,563)	HK$192,776

Source: *Guangzhou Yearbook 1999*, 1999; *Hong Kong Annual Report 1999: A Review of 1998*, 1999.

to as the waste deriving from domestic and commercial activities. In HK, MSW includes non-hazardous industrial waste and domestic and commercial wastes. The total and per capita municipal waste generation rates for HK and GZ in the last decade are given in Table 2. To compare likes with likes, non-hazardous industrial waste is excluded from the waste statistics of HK in Table 2 (see "The consequences of waste management," p. 202, for analysis).

Collection and collection costs

Specialized vehicles are used to collect and transport domestic waste and to reduce the adverse impacts on the road environment. In this regard, at the time of writing Hong Kong had a municipal waste collection fleet of 429 vehicles and a crew of 9,528 to handle 6,030 tonnes of domestic waste a day (FEHD, 2000). The domestic waste collection and transport costs borne by the two former municipal services departments[1] were reported at

Table 2. Per capita and annual waste generation of Hong Kong and Guangzhou

year	Tonnes per day		Per capita (kg/day)	
	Hong Kong (excludes industrial waste)	Guangzhou	Hong Kong (domestic only)	Guangzhou
1982	6500[@]	1001	n.a.	0.43
1990	5833	2163	1.00	0.81
1991	5957	2543	0.98	0.82
1992	6221	2702	1.01	0.82
1993	6571	2970	1.02	0.83
1994	6765	3076	1.04	0.85
1995	6730	3187	1.01	1.05
1996	7348	3625	1.00	1.09
1997	7975	3896	1.04	0.98
1998	8108	4555	1.00	1.14

@ Waste statistics in Hong Kong before 1983 were not thoroughly compiled and the figure quoted includes construction and industrial wastes.

Sources of HK data: Environmental Protection Agency, 1983; EPD, 1991, 1993, 1995, 1997a, 1997b and 1999a.

Sources of Guangzhou data: *Guangzhou Yearbook 1999*, 1999; Guangzhou Construction Committee and Guangzhou Environmental Health Bureau, 1999; pers. comm. with Z.H. Lei on 15 May 1999. Per capita waste figures are worked out by the authors.

HK$370/tonne in 1997 prices (EPD, 1999c). This does not include the cost of waste removal within the residential buildings and from the residential buildings to the refuse collection points. The capital costs of the refuse collection points and the refuse collection vehicle depots and their land costs are also excluded.

For the city of GZ, there is a RCV (Refuse Collection Vehicle) fleet consisting of 250 vehicles, each with 6–7 tonnes of carrying capacity and a frontline cleansing and waste collection crew of 8,871 to handle 4,555 tonnes of solid waste each day.[2] Collection costs, including intermediate transfer in GZ, was about RMB74.76/tonne of waste at 1995 prices (Guangzhou Construction Committee and Guangzhou Environmental Health Bureau, 1999, p. 41). Taking into account the inflation in wages in GZ, the authors estimate that collection costs were around RMB78.4/tonne in 1997.[3]

Waste composition

Figures 2a and 2b show the MSW composition of the two cities in recent years. Compared to HK, GZ has significantly less paper and metals but more putrescibles, wood and ashes (as reflected in the "others" category) in its disposal stream.

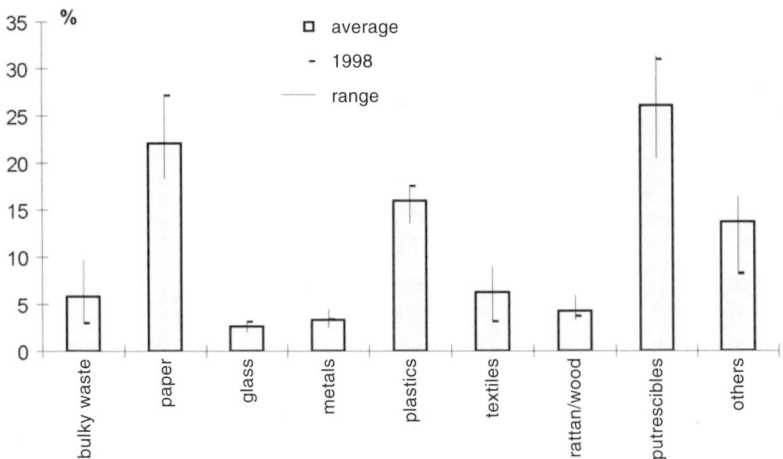

Figure 2a. The waste composition of HK, 1989–1998

Source of data: EPD, 1999b.

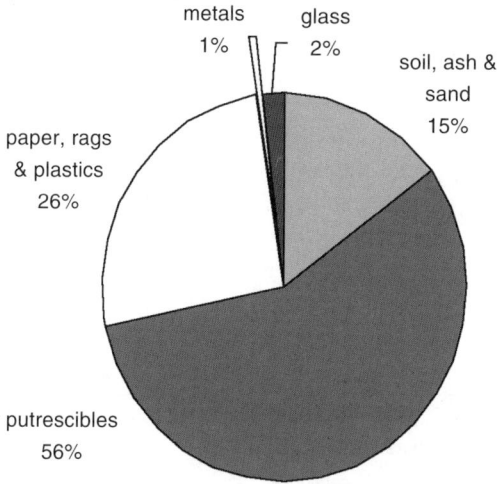

Figure 2b. The waste composition of GZ, 1999

Source of data for GZ: Chung and Poon, working paper.

Recycling and recyclable contents

The recycling rates and recyclable contents in the disposal waste streams are indicators of how recycling is being implemented in a society. According to the estimation of the concerned GZ authority, the overall waste recovery rate was about 40% with metals, waste paper and plastic scrap, reaching a recovery rate of over 50% (Guangzhou ershiyi shiji, 1998).[4] Over the past decades, about 30% of the MSW in HK has been recovered for recycling. Surveys also indicate that the recycling rate for industrial and commercial waste in HK may be as high as 53% of the generated quantity but that for household waste is only around 8% (Planning, Environment and Lands Bureau, 1998).

Recyclable contents surveys on the waste stream of HK indicate that a greater amount of recyclables have been found in recent years (see Table 3). For GZ, recyclable contents data in temporal series are not available. Recent data were obtained from an academic study. Figure 3 compares the latest recyclable contents of the domestic waste streams in GZ and HK. It shows that there are fewer recyclables in the domestic waste stream of GZ, except for wood, rags, and clear plastic bags.

Table 3. Recyclable contents in MSW of HK (1989–1997)

	Recyclable contents in domestic waste	Recyclable contents in commercial and industrial waste
1989	31.0%	31.9%
1990	26.9%	24.1%
1991	23.3%	20.7%
1992	24.9%	29.5%
1993	46.0%	40.3%
1994	55.7%	59.5%
1995	41.7%	51.8%
1996	n.a.	n.a.
1997	49.8%	47.6%

Sources: EPD, 1991, 1993, 1995, 1997a, 1997b, 1999a. No survey on the recyclable contents of waste was carried out in 1996.

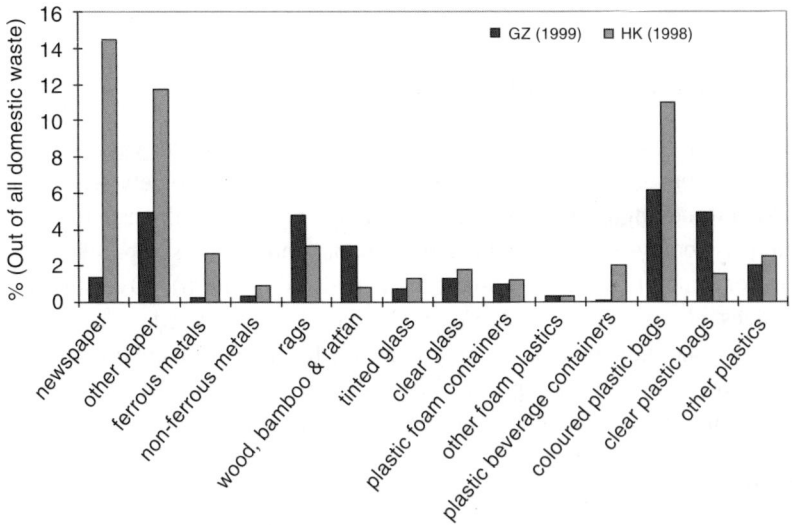

Figure 3. Recyclables contents in the domestic waste streams of GZ and HK

Source of data for Hong Kong: EPD, 1999a; Source of data for Guangzhou: Chun & Poon, Working paper.

Institutions

Waste management

In municipal Guangzhou, the management of MSW falls under the jurisdiction of the Municipal Construction Commission (MCC). The Environmental Health Bureau (EHB) is its major administrative organ in charge of the collection, transportation, and disposal of MSW as well as the related planning and the day-to-day operation of waste facilities at the municipal level. District EHBs were established to take care of the day-to-day waste collection and street cleansing work of the respective districts. The municipal EHB is responsible for monitoring the performance of the district EHBs. At the macro level, the municipal EHB also plays a major role in formulating waste policies. For the monitoring and control of the environmental effect of MSW management, it is the responsibility of the municipal Environmental Protection Bureau (EPB).

Frontline waste removal work, i.e., from the households to the neighborhood waste collection points, such as the "refuse huts", however is the task of various Street Management Offices (SMO; *jiedao banshiju*). In GZ, each SMO has many responsibilities, such as district security, family planning, political education and environmental hygiene campaigns, to cite a few, within its own "streets", which effectively form a small community. Performance monitoring and assessment of SMOs with respect to environmental hygiene, including frontline waste collection services, are carried out by the district EHBs.

The institutional structure for waste management in HK is different from that in GZ in that more private parties come into play at the operational levels. The responsibilities are also more centralized, with only one policy bureau and two departments involved. Waste management policies are formulated by the Environment and Food Bureau (EFB), formerly, the Planning, Environment and Lands Bureau (PELB). The Environmental Protection Department (EPD) is its executive arm in the implementation of waste policies, the formulation of waste management plans, and the planning of waste treatment, disposal and transfer facilities. The Food and Environmental Hygiene Department (FEHD) took over the waste collection and street cleansing duties of the Regional Services Department and the Urban Services Department after a restructuring of government institutions. The waste services provided by the FEHD include the running of district refuse collection points, transporting the waste from these points to the refuse transfer stations (RTSs) or other disposal facilities and street cleansing.

Frontline domestic waste collection and the collection and transportation of commercial and non-hazardous industrial waste are carried out by individual cleaners or private cleaning companies at the expense of the waste generators. Through contractual control, the management and operations of most large-scale waste facilities, such as landfills, RTSs and Chemical Waste Treatment Centres (CWTC), are entrusted to private hands, while the government assumes the role of performance monitoring.

The recycling sector

Though comparatively speaking, more government involvement in recycling is found in GZ than in HK, the private sector is more active than the public sector regarding waste recycling in both cities (Chung and Poon, 1998b).

In GZ, as well as in other parts of the mainland, waste recovery is the jurisdiction of the Supply and Sales system (*gongxiao xitong*). A number of state enterprises such as recyclable recovery firms (*wuzhi huishou gongsi*) and recycling plants were established in Guangzhou. In addition, a recyclable management office (*zaisheng ziyuan guanli bangongshi*) was formed to manage the waste recovery sector.

In HK, waste recovery is dominated by private enterprises that take care of the collection, reprocessing and exportation of recyclables. In recent years, some source separation initiatives in the form of "bring type" source separation programs (SSPs) have been noted by the government and are mainly administered by the FEHD and the public housing sector. In these "bring type" SSPs, recyclers are required to bring the source separated recyclables to recycling bins at public areas outside their buildings, such as municipal parks, building entrances, commuting points, etc. But there is no governmental involvement in the marketing and reprocessing of materials as noted in the case of GZ.

Comparative analysis of the waste management systems in GZ and HK

The waste management systems of GZ and HK will be analysed in terms of their policy ideologies, policy contents, policy processes, public participation and policy consequences. This analytical framework, modified from

Hoberg's ideas (1986), has been used on previous occasions (Lo and Chung, 2000):

- policy ideology considers the policy priority of the municipal government and social groups with regards to sound and effective waste management and attitudes toward the idea of proactive control of wastes.
- policy content refers to the use of legislative and policy instruments for waste management.
- regulatory process refers to the structures of key institutions and the implementation of waste policies and regulations.
- public participation considers the participation of social and non-governmental forces in MSW management. The extent and enhancement of environmental citizenship in waste management is analyzed.
- policy consequences consider the outcomes of municipal waste management.

Traditionally, analysis and discussion of waste management has been oriented towards various operational aspects: waste collection, transportation, pre-disposal treatment, disposal, recycling and reduction etc. A modified Hoberg framework offers an alternative perspective that is comprehensive enough to capture all the important aspects of waste management and allows for a critical investigation into the institutional features of waste management systems. To enable an understanding of the subsequent comparative analysis, readers are encouraged to refer to other literature on the two waste management systems (Chung and Poon, 1998a, 1998b; Lo and Chung, 2000).

Waste management ideologies in GZ and HK

Tracing the policy ideologies in waste management in HK goes back to the 1980s. The Waste Disposal Ordinance of 1980, the Waste Disposal Plan of 1989, the environmental policy paper of 1989 and the early "Keep HK Clean" Campaign all focused on clean and speedy collection of all solid waste, a reduction of littering, and cost-effective disposal and waste transfer services (PELB, 1989). In short, the aims of earlier waste management policies were to provide low cost and efficient end-of-pipe treatment for the waste of HK.

By 1998 the Waste Reduction Framework Plan (WRFP) was launched.

It explicitly adopts the waste management hierarchy and proposes a series of waste reduction and recycling policies and measures. Although the WRFP appears to pay more attention to higher order waste management needs than previous policy plans, most waste reduction measures in the WRFP are either in the pipeline or have not been implemented long enough to generate an impact. In reality, the legacy of the former waste management ideology as set out in the Waste Disposal Plan of 1989, oriented towards satisfying lower order needs in waste management, is still highly conspicuous.

In GZ, there was a lack of a clear and specific waste management ideology in the early years. With the promulgation the 1995 *Zhonghua renmin gongheguo guti feiwu wuran huanjing fangzhifa* (National Law on Solid Waste Pollution and Its Prevention of the People's Republic of China), the ultimate objectives for waste management, namely, safe disposal, recycling and reduction, were defined. Despite the 1995 National Waste Law, the need to address waste management needs was not formally on the administrative agenda in GZ until 1998 upon the release of the *Agenda 21 for Guangzhou*. In this policy plan for the coming century, the waste management hierarchy was acknowledged as in the case of the WRFP in HK, though in a more implicit way. Yet, the prospects for a proactive approach are limited due to the lack of any clear and aggressive policy toward waste avoidance, recycling and even safe disposal (see "Policy contents in waste management", p. 193).

This seems to suggest that in both cities, despite an awareness of and a policy intent to put higher order waste management measures in place, empirical evidence shows otherwise.

Does the public in the two cities support the waste management hierarchy as well? The findings from several recent popular surveys and field studies tend to suggest mixed sentiments in both GZ and HK. On one hand, popular surveys have shown that most people in Guangzhou are familiar with the concepts of conserving resources and recycling and regard them as praise-worthy behaviour. As for a commitment to proactive resource conservation actions, the findings also indicate that an overwhelming majority of Guangzhou citizens have habitually recovered their own recyclables. However, the same survey also suggests that resource conservation behaviour is associated with the expected monetary returns from redeeming the recyclables.[5] Findings from a number of environmental and waste recycling attitude surveys in HK suggest similar verbal support for waste recycling and reduction. However, when compared to their GZ counterparts, HK people are less keen to put their beliefs into practice on a regular basis.[6]

Simultaneously, other field surveys indicate that despite general agreement on the environmental merits of resource conservation, rapid economic development has given rise to more wasteful lifestyles among people in both cities. Studies have found that the proportion of recyclables in the disposal waste streams in GZ and HK has increased in recent years (see Table 3, and Chung and Poon, 1998b). The seemingly conflicting observations — substantial verbal support for resource conservation on one hand but more wasteful lifestyles on the other — suggest a low level of public commitment to consumer-oriented sustainable resource use.

On the whole, one may conclude that the popular view in both cities on waste management has gone beyond the narrow horizon of environmental health and sanitation to a non-committed sustainability orientation. In the case of GZ, the incentive for popular waste recycling support is likely to be based on economic considerations, a drive that can be effectively taken advantage of, if so wished, to further higher order waste management performance.

In short, both the waste authorities and the general public in GZ and HK are in a state of transition, changing from satisfaction with the simple fulfillment of lower order needs in waste management to demands for good performance in waste prevention.

Policy contents in waste management

To what extent are the waste management ideologies being translated into the policies of the two cities and are they being translated in a coherent and integrated form necessary for meeting both higher and lower order waste management needs?

Improvements in collection are by far the major accomplishment in MSW management in GZ. GZ has almost finished the upgrading of frontline waste collection services by replacing the conventional method with door-to-door bag collection services.[7] Associated with such an improvement is an increase in the waste removal and cleaning fee charged by the SMOs, from RMB4/household/month (hh/mth) to RMB10/hh/mth. Currently, the management of solid waste from collection to disposal in Guangzhou is strictly a heavily subsidized government responsibility. Only a small part of the operation costs is recovered by charging individual waste generators (e.g., the RMB10/hh/mth; see also "Cost effectiveness," p. 210). Realizing that the traditional burden to provide welfare to public servants cannot achieve high efficiency in labour-intensive tasks, the waste management body in GZ has a plan to contract out waste collection and street cleansing services.[8] However, the market sector for waste collection and disposal is so underdeveloped that the prospect of using market instruments in MSW management will be limited for the coming future.

The safe disposal of waste is still a goal to be attained in GZ. Only about 63% of solid waste was considered safely disposed of at the two landfills in GZ.[9] Such an unsatisfactory situation will continue for another 2–3 years until the replacement landfill comes into full operation.

Waste avoidance, though recognized as a target (Guangzhou ershiyi shiji, 1998, p. 153) is a neglected area in GZ. Waste recycling is another higher order waste management measure that enjoys greater attention.

A government-sponsored waste separation pilot program was launched by the EHB in conjunction with the SMOs in 1998 whereby residents are requested to discard separately dry recyclables, namely, paper, aluminum cans, glass, rubber, plastics, and metals[10] (see *Nanfang ribao*, 1998, p. 1; *Guangzhou ribao*, 1999). At the time of reporting, about 40,000 households had already taken part in the extended phase of the waste separation pilot program.[11] Though SSPs are expanding in scale, there is no challenging recycling target to keep up the momentum.[12]

A reason for the ineffective performance of GZ's solid waste management is that in the absence of local legislation on solid waste management, responsible agencies lack a solid legal basis for prosecuting violations. The National Waste Law of 1995 and the municipal "Regulations on City Appearance and Environmental Hygiene Management of Guangzhou" in 1996 are crucial steps in the creation of a legal framework for waste management in Guangzhou. However, their contributions are limited as the law, which sets the broad policy directions and provides an administrative scheme for waste management, is too generic for local consumption, and the regulations are concerned almost exclusively with city cleansing, the prohibition of littering and the definition of waste collection responsibility. In sum, there are some commitments to improve lower order waste management measures but policy measures to attain higher order waste management needs are limited in scope, and they lack robustness.

Relative to GZ, lower order needs of waste management are much better taken care of in HK. Since the strategies in the Waste Disposal Plan were implemented in 1989, HK has been able to establish satisfactory physical networks and management procedures for municipal waste collection, storage, transfer, disposal and treatment.

Contrary to GZ, most of the authority in environmental administration comes from local laws. The Waste Disposal Ordinance (WDO) sets the statutory framework for MSW management in HK. Waste laws in HK are strong at meeting lower order needs but weak in providing initiatives for higher order measures. For instance, the Waste Disposal Ordinance requires the concerned policy secretary to plan for the collection and disposal of solid waste without due consideration to the need to reduce and recycle waste. Instead, higher order waste management requirements are promulgated in policy plans, policy papers or administrative rules that lack comprehensiveness and force.

In the WRFP, a 40% reduction target on the quantity of waste for disposal shall be achieved by the year 2007. Half of the reduction target will be achieved with volume reduction techniques through waste incineration and composting. The other half is to be achieved through a range of preventive measures, such as voluntary community waste separation schemes, sectoral scheme and environmentally responsible purchasing, land allocation schemes, producers' responsibility schemes, material recovery/recycling facilities and economic measures, namely charging for waste disposal. However, despite the multi-pronged approach of the WRFP, there is a lack of measures to systematically transform HK into a resource conserving and recycling oriented society.

Although charging for waste disposal has been made into a policy, domestic waste collection, refuse transfers[13] and landfilling are still free government services. Industrial and commercial establishments in Hong Kong only have to bear the cost of waste delivery up to the point of the refuse transfer stations or landfill disposal. The tariff structure of private waste collectors is usually set at a proportional or even regressive rate. Similar to the case of GZ, domestic waste generators have to pay for the removal of their waste. But unlike the case in GZ where a separate waste collection and cleansing fee is collected, the waste management costs to a HK household are imperceptibly included in the property management fee and bear no relation to the amount of waste generated. It is therefore not surprising that most people do not know how much they are paying for waste management.[14] Thus there is very limited scope where economic instruments can be used to generate incentives to reduce waste in the present waste charging structure.

Even if a waste disposal charge were to be introduced, at its present proposed form the charge will only be set at 50% of the capital and operation costs of landfilling, which amounts to about HK$62.5/tonne of waste at 1999 prices. No charges will be incurred for domestic waste disposal. The rate itself is set neither at a cost-recovery level nor at an incentive level. For one, landfill replacement costs, environmental costs and the use of intermediate waste facilities are not included (EPD, 2000). For another, the effect of a low HK$62.5/tonne fee is unlikely to motivate a normal commercial and industrial waste generator to reduce waste. In 1997, the average generation rate was 1.1kg/day/employee (EPD, 1999a). For a medium sized company with 100 employees, daily waste generation is expected to be around 1.1 tonnes. Therefore, the total proposed cost of waste disposal is a mere HK$68.8 per day, a negligible amount to any profit-making company. As such, it is only be fair to consider the charge, even if imposed, at its best merely a preliminary step towards waste reduction in HK.

Similar to the case in GZ, SSPs are seen as key measures in promoting waste recycling. The HK government's encouragement of community SSPs has resulted in a proliferation of participating housing estates; there were 227 in September 1999 covering 787,366 households (Waste Reduction Committee, 1999, p. 4). However, the sheer growth in the program coverage has not been matched by similar improvements in waste separation. A survey conducted in April 1999 revealed that community SSPs accounted for less than 30% of the collected household recyclables[15] (Centre of Environmental Technology, 1999, p. 14). Also, in a randomly sampled survey carried out in 1998, it was found that only 14% of the respondents had recycled through SSPs (Chung and Poon, 2000). In terms of the quantities recycled, the government-led SSPs in total collect about 3800 tonnes per month (Waste Reduction Committee, 2000) or less than 2% of the domestic waste disposed in 1998.

In short, in both cities, even though policy measures to cater to lower order needs in waste management have been instituted, albeit to varying degrees, policies to meet higher order needs are far from sufficient and both legislative and economic support are less than adequate.

Regulatory processes of waste management

The regulatory processes of waste management in GZ and HK are both marked by separate jurisdictions albeit to different degrees. The GZ system is dominated by a command and control approach, while the system in HK stresses informal collaboration between regulator and regulatees to achieve voluntary compliance by the latter and cooperation with the former, namely the EPD which is the chief regulatory agency with a wide range of enforcement power.

The GZ system features separate jurisdictions with limited coordination between the three main regulatory agencies: the EPB, the municipal and district EHBs and the SMOs. In this multi-tiered structure, communications and interactions between parties are usually in the form of performance assessments and meetings. The adequacy and frequency of communications between the interacting bodies depend on the individual officials' commitment and goodwill.

Accountability in this multi-tiered structure is weak, with no incentives for improvement. The SMOs derive their operation budgets from the RMB10/hh/mth waste collection fee. But the SMOs are in charge of a large number of community matters and this revenue from waste collection charges is not earmarked for waste collection and cleansing services. In operations, the district EHBs have to be accountable to their respective district governments and to the municipal EHB. However, the recurrent budgets of the district EHBs are allocated by the respective district governments, which in turn derive their revenue from the tax money in their respective districts. The separation between the sources of funding and the sources of duty assignment limits the capability of district EHBs to meet the requested performance targets of the municipal EHBs.[16]

Limited interaction owing to the separation of jurisdiction in waste management is also exemplified in the waste recovery sector of GZ. In Guangzhou, the waste recovery sector is regulated by the Guangzhou Recyclable Management Office and the Public Security Bureau. The former is responsible for reflecting the recyclable collectors' views of the government and for issuing licenses to the private wastepaper collectors. The Public Security Bureau monitors illegal trading of swags through the waste depots. There is a lack of formal communication between the ultimate authority in waste management (i.e., the EPB), the EHB and these two agencies. As such, the recent municipal EPB's decision to ban non-degradable food containers[17] has attracted criticism from the Guangzhou Recyclable Management Office that the ban discriminates against the waste recovery sector and is not heading in the right direction.[18]

In HK, at the level of policy execution the EPD is the chief authority for waste management. It has to interact with a number of government bodies in the formulation of measures and day-to-day waste management work. While the jurisdictions concerning waste management are relatively well-defined and centralized in one party, discordant intents in policy and measures can be found.

In actual day-to-day management, the EPD interacts with the waste collection authorities (FEHD, or formerly the two municipal services departments and the two municipal councils) that are responsible for the collection of domestic waste. The adverse impact from the separation in the duty to collect and the duty to dispose was not substantial until the change in waste management ideology to promote waste recovery. The three authorities, the EPD, the Urban Councils and the Regional Councils, are charged with different and at times incompatible missions. On one hand, the chief party, the EPD, realized that there is a waste disposal facility crisis. On the other, the waste collection authorities were not keen to make a change (i.e., to separately collect waste and recyclables) to assist the EPD in promoting recycling activities in Hong Kong. The lack of support from the public agency can be blamed for the poor performance of the SSPs that were set up before 1998; they were sporadic, ephemeral and lowly profiled. In these SSPs the collection of recyclables, financing and other logistics were the responsibilities of the program organizers.[19] However, it should be remembered that the collection, transfer and disposal of MSW is almost fully-subsidized by the government. However, if there were no SSPs, the materials would have been collected by the government out of the taxpayers' pockets.

Having said that, signs of possible improvement are evident. With the restructuring of the municipal and environmental institutions, both the EPD and the newly formed FEHD were brought under the EFB. It is possible that better coordination and communication will appear in the future. However, further improvements to meet higher order waste management needs may face greater resistance from other sectors and even within the government. An example is the land allocation policy which says that government land will be leased to the private waste recovery sector at a reasonable cost. Two years after its introduction,[20] the policy has met with considerable criticism. Other than the dissatisfaction of outsiders to the policy,[21] internal support from respective government bodies (such as the Lands Department) and popular bodies such as the District Boards is lacking. This makes the acquisition of sites for the recycling trade a lengthy and difficult process, with the majority of sites being rejected for use in recycling operations (Waste Reduction Committee, 2000).

In sum, when compared to GZ, solid waste management duties are less separated in HK, resulting in fewer co-ordination problems than in a separated jurisdiction system concerning waste management basics. The lack of accountability in the waste management authorities in GZ is due mainly to the fact that disparate parties assume the budget granting role and the performance monitoring role. In HK, however, there is a lack of a thorough understanding of the urgency to meet higher order waste management needs in the public sector and popularly elected bodies, thus rendering well-meaning policies ineffective (see "Waste management ideologies in GZ and HK," p. 191).

Participation in waste management by the public

The nine components of environmental citizenship, namely, information, awareness, concern, attitudes/beliefs, education and training, knowledge skills, literacy and responsible behaviour (UNESCO, 1978; UNESCO, 1993; Hawthorne and Alabaster, 1999) are interrelated concepts that can be reorganized into five groups, literacy/knowledge, awareness, skills, education/training and responsible behavior. These five groups have a bearing on MSW. In the present discussion, responsible behaviour shall include practicing consumer-based sustainable waste management, such as recycling and waste avoidance as well as scrutiny of the waste authorities.

In Guangzhou, formal public participation is not a component part of the waste management system. Neither the decision making and regulatory processes are open to external forces. The system is dominated by the Guangzhou Construction Committee (that heads the EHB) and the EPB without any formal channels for public consultation and citizen participation. All solid waste policies and regulations are undertaken by these two bureaucratic agencies in the absence of public consultation and supervision. The regulatory process operates more or less in a 'black box' manner, with a very low degree of transparency. No open consultation is allowed for people to express their opinions, nor for those affected to register their grievances, even in cases where MSW management has profound impacts on the citizens. In addition, public accessibility to waste management information and statistics is low. The only point where public views may have some weight is at a micro-level through direct communication with the SMOs, whose services are directly collected from the households. The lack of genuine public scrutiny has greatly reduced the accountability of the waste management system. As a whole, the existing institutions generally do not safeguard or maximize the public interest which is likely to be neglected in favor of bureaucratic and business interests.

The municipal EHB has recently taken some positive steps to encourage environmental citizenship on waste issues. Opinion surveys have been conducted to collect people's views in the formulation of solid waste policies. For example, an opinion survey on domestic waste sorting was conducted by the Guangzhou Environmental Health Institute in October and November 1998 to determine the degree of popular support for the new initiative for waste sorting programs. At the same time, citizens are encouraged to lodge complaints against solid waste pollution and waste management services to the municipal government and its waste management agencies through the 'Mayor's Hotline' and specific complaint hotlines located at every refuse collection depot and transfer station. The improvement in the people's living standards as a result of rapid economic development has also made the general public more aware of the deteriorating quality of their living environment. In addition, people are now more informed of different pollution problems as the local mass media have indicated a growing interest in reporting on local and national environmental issues. Thus people have become more outspoken on pollution problems and the mass media have provided them with a channel to voice their environmental concerns. However, awareness, attitudes and the theoretical literacy of the average GZ citizen on sustainability (resource management) are still low (see "Insufficient public consciousness on sustainable waste management," p. 214). In addition, since there is no party to act as an ombudsman within the environmental health system, even if complaints are lodged at the right office, it is unclear to outsiders as to how effectively the rectification will be carried out.

In HK, the problems and issues regarding public participation in waste management are different from those in GZ. Public involvement in HK is a regular feature at both the policy and operation levels, with the existence of elected bodies, advisory committees, and non-governmental environmental groups. Institutional channels for public consultation include the District Boards, the two municipal councils and the Advisory Council of Environment. To some extent public scrutiny is carried out by professionals and interest groups in these institutions as well as by the municipal councils and district boards. Concurrent with this institutional development for facilitating social participation, the public is now more informed about waste management. Environmental information, whether technical, educational or management oriented, can be accessed in a variety of ways: government web-sites, annual reports of the relevant government departments, environmental resource libraries, hotlines or direct enquiries.

In addition, major policies or measures are usually released for public consultation before they are enacted. Thus, it appears that HK is ahead of GZ in all respects in terms of environmental citizenship regarding waste issues.

Yet, having the opportunity to voice opinion does not guarantee influence. At times, public views were so influential that they were cited as the main reason for delaying some policies, such as with the waste disposal charge. Yet, public opinion can still be ignored. A case in point was the public consultation related to the 1989 Waste Disposal Plan. The 1989 Plan contained outdated ideas even at the time it was released for public consultation. During the consultation, community groups pointed out the landfill crisis and advocated a shift of emphasis from disposal to waste reduction and recycling (Conservancy Association, 1989). Yet, this foresightedness was not incorporated into the final Plan.

Public demand for higher order waste management needs have continued to be heard over the past decade. In a number of surveys the public has revealed its support for a convenient household waste recycling infrastructure and demands have been increasing. Yet, not until 1998 did the public housing estates start to offer more permanent SSPs which nevertheless belong to the less-convenient bring system instead of the more popular door-to-door collection system.[22] More importantly, these recycling programs are not incorporated and regularized in the municipal services system, making them appear as trial runs. This also causes a vicious circle: poor SSP logistics leads to a poor recycling program which gives the impression that the public does not care to recycle (see "Policy contents in waste management," p. 193).

Thus, despite all the consultation mechanisms in Hong Kong and the high accessibility of information, regular public input is mainly limited to those groups having representation on the statutory committees. Opportunities for participation by the masses are only offered irregularly. The public at large remains passive and reactive. More importantly, the receptivity of the public agency to public opinion and the public demand for higher waste management needs is limited, thus inhibiting citizens' responsible behaviour for sustainable waste management.

The consequences of waste management

This section will analyze the policy outcomes of waste management arising from the policy intent and regulatory behaviour of the authorities as well as from the regulatees' reactions. Four issues are discussed as outcomes of MSW management: collection, management and use of disposal facilities, generation and recovery. To conform to the waste management hierarchy, a waste management system should be able to provide a low waste generation level, high waste recycling rates and minimal environmental damage (including ecological damage) from waste disposal and collection.

Waste collection

Waste collection is very backward in GZ and the resulting pollution is obvious. Although the collection rate is high when compared to some cities in developing countries, the majority of Guangzhou citizens considers environmental hygiene in the city to be unsatisfactory. This is supported by an opinion survey conducted by the Guangzhou Environmental Health Institute and the Guangzhou Public Opinion Research Centre in 1998 which indicated that only 13.4% of the respondents rated sanitation conditions as 'good' or 'quite good' (1998, p. 2). Furthermore, in a self-evaluation exercise, the adequacy of MSW collection facilities, waste collection coverage and performance in achieving a hygienic environment during and after waste collection were graded as merely acceptable or not acceptable.[23]

On the surface, all aspects of MSW collection in HK are being taken care of. However, backward waste collection facilities can still be found. In contrary to the better equipped permanent refuse collection points,[24] temporary refuse collection points are sometimes simply open bin sites. Thus, it is not surprising that from 1996 to early 1999 the number of complaints about refuse collection, littering and cleansing received by the two municipal services departments increased from 2,602 to 4,218 cases (Urban Services Department, 1997, p. 37; Provisional Urban Services

Department 1998, 1999b, p. 42; Regional Council, 1998, 1999, p. 38).[25] This can be translated to about 11.6 complaints per day or more than one complaint in each administrative district every two days.

In sum, collection has not been properly carried out in GZ, resulting in poor hygiene. HK performs better than GZ in most aspects of waste collection although there is still room for improvement in terms of the collection, handling and facility provision of MSW.

Management and use of disposal facilities

Although they are the major waste disposal facilities, both landfills in GZ are poorly managed, falling short of the standards for sanitary landfills (Guangzhou shi renmin zhengfu, 1996, p. 74). Although pollution control measures, such as an impermeable lining, gas recovery and leachate treatment facilities, are used in the landfills, they are inadequate to safeguard against pollution in the surroundings. Worse still, the laggard process for the siting and planning of new waste disposal facilities has caused an undesirable extension of the service periods, further overloading the already inadequate leachate collection and treatment facilities.

In addition, site management is poorly conducted. Waste compaction is not carried out by landfill compactors but by earth-movers. There are no clear space limitations of waste loading areas. As a result, pollution into the nearby agricultural environment has been widely noted. It was reported that the BOD and COD concentrations of the treated leachate from both Da Tian Shan and Li Keng exceeded the city's effluent standards by more than 20 times and 10 times respectively (Lu, 1997). Furthermore, the disorderly scavenging practices of the waste miners have often obstructed the timely off-loading of waste and soil covering in the landfill proper. At the same time, it has also caused injuries to the waste scavengers.

The costs and benefits of waste management are hardly satisfactory in Guangzhou. Under the current command and control approach, the municipal government has to finance waste management in Guangzhou. It has to grant a budget to both the EHB and EPB to manage municipal and other wastes. The cost for MSW management is currently estimated at about RMB120/tonne at 1995 prices (Guangzhou Construction Committee and Guangzhou Environmental Health Bureau, 1999, p. 41, Table 5-1). In order to attain reasonable standards of pollution control in landfill sites and to safeguard adjacent fresh water resources, future landfills in Guangzhou should be better equipped with pollution control facilities. It has been estimated that the cost for landfilling and waste collection will be RMB200/ tonne of waste, an increase of over 66% (Guangzhou Construction Committee and Guangzhou Environmental Health Bureau, 1999, p. 42). With the expected continuous increase in waste generation, total waste management expenditures in Guangzhou will further increase. This is a heavy burden on the municipal government, despite the city's fast growing economy.

Major MSW facilities in HK, including refuse transfer stations (RTSs) and landfills, have largely been under appropriate management without which they would be a source of environmental pollution. The RTSs, serving as the main intermediate waste facilities for temporary storage, are equipped with a water treatment system to reduce the pollution loading of the surface drainage and sewage from the waste containerization process. As these RTSs are generally distant from sensitive receivers, odor and noise nuisance on local communities is low. The landfills in Hong Kong are equipped with pollution control devices, including double layers of impermeable liners, gas venting systems and leachate collection and treatment work. Site management is also carried out to ensure the speedy unloading of waste matters.[26] This is coupled with extensive monitoring of all water, air and noise emissions from the site (EPD, 1995a,1998a). In short, GZ lags behind its HK counterparts in waste facility management for MSW.

Waste generation

Statistics over the last few years have indicated that the GZ MSW system has not been able to control the rapid generation of waste. Table 2 shown earlier indicates that the per capita MSW generation rate increased from 0.81kg/day in 1990 to 1.14 kg/day in 1998, an increase of almost 41%. When one considers the absolute quantity of waste generated for disposal, the increment of 2,392 tonnes per day was even more threatening as this represented an almost 111% increase in the need to improve waste collection and disposal in order to cope with the by-products of the economic boom. Waste generation statistics in HK show similar trends, although the increase is mostly to the total quantity (i.e., due to population increase) and not to per capita generation (see Table 2).

In sum, in both GZ and HK the sustainability requirement of waste reduction is not met. This runs counter to the ideological preference of the waste management hierarchy.

Waste recovery performance and recyclable contents

Recovery rates and the recyclable content of the waste stream are also measuring rods of the success or failure of the waste recovery and source reduction measures.

Most recovery activities in GZ have been carried out by the private sector through an informal system. The informal nature of these waste recovery activities is best illustrated by the bottom tier and the frontline waste recovery parties of the system — the cleaning crew of the sanitation stations, the scavengers, and residents. However, the municipal cleaning crew is prohibited by an administrative order from recovering recyclables while they are on duty,[27] meaning that waste is recovered only when the crew is off duty or the waste is not checked. At the same time, scavengers and residents recover waste for reselling at recyclable depots in the neighborhoods in a spontaneous manner motivated in most cases by the small but non-trivial economic return. Although there are state-run recyclable depots, the trading volume through the state network is estimated at only one-third of the total (see Chung and Poon, 1998b). Thus, most of the recyclables recovered are actually handled by the privately run waste depots and recyclers. However, it is difficult to evaluate the actual performance of the Guangzhou waste recovery system due to the lack of comprehensive and reliable comparative waste recycling data.

Despite the intent to tackle higher waste management needs in the WRFP, waste recovery activities in HK are still largely carried out by the private sector that in the past decade has been diverting over one-third of the MSW from the disposal stream (PELB, 1998). There are two main differences between private waste recovery activities in GZ and HK. First, source separation of waste by the general public in HK is seldom driven by economic motives as it is in the case of GZ. Second, recovery activities in HK are mostly concentrated on pre-consumer recyclables. Post consumer recovery is still at its infancy in HK while it is already a widely spread phenomenon in GZ (Chung and Poon, 1998b). HK statistics further show that total waste recovery[28] decreased since reaching a peak in 1995 (Table 4) while total MSW generation increased during the same period.

This naturally leads to the outcome that recyclable contents in the disposal streams of HK are increasing (see Table 3 and Figure 2). However, HK is not alone in performing poorly in waste recovery. Despite the lack of data for recyclable contents for earlier years in GZ, a review of the waste

Table 4. Waste exports and recycling in HK (1993–1998)

	Exported and locally recycled (tpd)
1993	4848
1994	5139
1995	5296
1996	4479
1997	4224
1998	4273

Sources: EPD, 1995, 1997a, 1997b, 1999a, 1999c.

composition change over the past few years also suggests that recyclables (made up of paper, plastic, metals, glass and rubber) in its waste stream have increased from approximately 6.6% in 1985 to 31.2% in 1999 (see Chung and Poon, 1998b; Chung and Poon, working paper).

In sum, despite the laudable intent to reduce waste and to encourage MSW recycling in both cities, the reality is that both HK and GZ have moved further away from a low waste and conserving society (Table 5).

Evaluation of the waste management systems in GZ and HK

To what extent are the waste management needs specified in the waste management hierarchy met in HK and GZ? Four criteria will be used to carry out an evaluation.

Environmental impacts

An evaluation of the environmental impact of an MSW system examines whether waste matters are tackled by the best available technical means throughout their life cycles (i.e., from consumption, collection, transportation to treatment). The objectives are to reduce the direct pollution impact and the less direct resource depletion problems.

It is clear from the previous analysis that both systems fall short of practising low environmental impact waste management. The waste management system in Guangzhou has created considerable adverse environmental impacts in the course of MSW management and itself has become a source of environmental pollution (Guangzhou shi renmin zhengfu, 1996, p. 73). The use of garbage tanks in open areas as the major

Table 5. Summary of the features of the two waste management systems

	GZ	HK
Policy ideology		
Of the authority:	• Waste management hierarchy is implicitly adopted	• Waste management hierarchy explicitly adopted
Of the public:	• Support noted but is founded on economic gains	• Support remains at verbal level
Overall:	• Stepping up both lower and higher order waste management measures simultaneously	• Transforming from a lower order waste management orientation to a higher order orientation
Policy content	• Remedial, ineffective, piecemeal • Under use of economic incentives	• Over-reliance on voluntary and administrative measures and lack of an integrated and forceful plan to institute higher order waste management measures • Under use of economic incentives
Regulatory process	• Command and control oriented • Separation of jurisdiction leads to a lack of accountability in the system	• Voluntary compliance • Discordant policy intents between collection, disposal and other authorities
Public participation		
Information accessibility:	• Low	• High
Literacy/ knowledge, awareness, skills, education/ training:	• Except for source separation skills, unsatisfactory in most aspects	• Slightly higher than GZ in terms of awareness but equally unsatisfactory in other aspects
Responsible behaviour	• Limited public scrutiny • Other than economically driven source recovery behavior, other responsible behavior is lacking.	• Moderate public scrutiny • Lacking in most responsible behavior
Policy consequences	• Poor collection • Lack of proper disposal facilities • Failure to limit waste growth • Lack of waste avoidance	• Generally satisfactory • Intermediate and disposal facilities are generally well-managed • Failure to limit waste growth • Lack of waste avoidance

means of waste collection has generated offensive odors and resulted in poor environmental hygiene. The substandard collection service has polluted the environment and generated excessive nuisances in the course of solid waste collection and transportation.

Furthermore, the capacities of most waste disposal and treatment facilities in GZ are limited, not to mention that these facilities are substandard and not properly managed. The lack of proper pollution control facilities in the landfills has compromised the safe disposal of solid waste and led to the contamination of water sources, air and soil in the areas as a result of pollution from leachate and landfill gas migration.

Despite some organized efforts to promote source separation of waste and the existence of an extensive waste recovery system in GZ, the prospect that GZ will become a resource conserving society is low, as indicated by the increasing amount of recyclables in the MSW stream.

In HK, the waste management system has been able to keep direct adverse environmental impacts from solid waste management at an acceptable level, while the capacities of the facilities are basically adequate to meet demand in the short to medium terms. However, the system's disposal capacity has been declining and will soon give rise to a landfill crisis — a result of the lack of diversity in waste disposal alternatives attributable to an over-reliance on a land-intensive disposal approach in the past.

In the case of recycled waste, there is also the need to ask if it is being recycled by proper technologies and checked by proper pollution control measures. In the case of HK, since the majority of the waste recovered is exported, and mainly to mainland China, comprehensive information on the level of technologies and on environmental controls employed in these plants is difficult to obtain due to a lack of surveys and studies on this aspect. In a number of reports, paper making, including paper recycling plants, was cited as a major polluter of water courses (Wang, 1999; Jiang, 1996; Faming and Wu, 1997). It may therefore be justified to cast doubts on the environmental integrity of recycling activities in GZ and thus some of the recycling activities associated with the HK secondary materials exported to GZ for recycling. Therefore, even though the direct environmental impact from waste management is adequately managed in HK, the indirect environmental impacts, namely the depletion of resources and the pollution from recycling associated with the secondary materials from HK, are expected to be high.

Implementation effectiveness

In GZ, the lack of effective implementation of waste policies and related regulations is the main cause of the poor performance in preventing environmental impairment from waste. Because of limited enforcement, legally prohibited behaviour such as littering, fly tipping, dodging the payment of waste collection fees,[29] and selling and using non-degradable food containers can commonly be found.

Examples of implementation failures in waste management are not lacking. The system has failed to control the rapid increase in the generation of solid waste and the source of white pollution, namely plastic waste. While there is a clear shortage of waste disposal facilities, planning for future facility needs is low.

The outright ban on non-biodegradable foam plastic food containers to encourage the use of biodegradable containers is another example of implementation failure. The requirement for the use of a different food container does not in itself provide an incentive for the public to reduce littering. The bio-degradable requirement is also inadequately formulated to make a significant difference in alleviating the ecological problems caused by non-biodegradable plastics. In fact, owing to the higher cost of degradable Expanded Polystryrene (EPS) containers, non-biodegradable EPS containers are still commonly found in the marketplace even though the ban has been implemented for one year (Zhao, 1998). In short, the ban has neither been able to achieve its intended environmental merits, nor has it been able to solve the environmental problems found in the waste management of GZ.

In HK, despite the success in containing direct pollution from waste management, the waste management institutions are not so capable of achieving higher order environmental needs in managing waste, such as resource conservation; little has been done or planned for waste avoidance and the government is still unable to provide economic incentives to encourage source separation for the promotion of waste recycling, namely to extend waste disposal charges to cover domestic waste. Even slower in progress has been the setting up of user friendly and popular systems to collect source separated materials.

Weak implementation of waste policy and limited enforcement of regulations have been found to various degrees in the entire process of waste management and have discounted the effectiveness of the regulatory control of waste. An example is MSW collection. The rapid increase in the number of complaints concerning refuse collection, littering and cleansing mentioned earlier tends to suggest that the performance of local waste collection services is not completely satisfactory. It appears that odor problems in refuse collection points and the accumulation of waste along roadsides in areas with high retailing activities are the two most conspicuous problems. From the figures in Table 6 which compare the ratios of refuse collection vehicles and cleansing staff in Hong Kong and Guangzhou, it is apparent the Hong Kong system has a slightly higher provision of RCV but lower availability of waste collection and public cleansing staff than GZ. Thus, despite the existence of a large refuse collection vehicle fleet, the provision or, more importantly, the efficiency of the cleansing staff should be raised to provide quality lower order waste management services.

Cost effectiveness

The cost effectiveness in solid waste management is generally satisfactory in Hong Kong. This condition can be illustrated by comparing it with the cost performance of waste management systems in different jurisdictions. Table 7 shows the relative cost of waste management in HK, Australia and GZ. The costs are expressed as percentages of monthly per capita GDP. The Australian data are included to offer reference to a developed country. The table shows that the relative costs for landfilling, domestic waste collection and capital investment for landfills in Guangzhou are higher than those in Hong Kong, while those in Hong Kong and Australia are similar. However, it should be noted that the landfills in Hong Kong are managed at a higher pollution control standard than those in Australia and

Table 6. Refuse collection vehicles and cleansing staff in Hong Kong and Guangzhou

	Hong Kong	Guangzhou
Vehicle per 10,000 people served	0.68	0.63
Vehicle per tonne of waste/day	0.065	0.064
No. of waste collection and cleansing staff per 10,000 people served	14.65	22.4
No. of waste collection and cleansing staff/tonne waste/day	1.41	2.27

The 1997 demographic and waste data of each city were used in the derivation. Sources of the data for Hong Kong: Provisional Regional Council, 1999; Provisional Urban Services Department, 1999. Sources of the data for Guangzhou: see Note 2.

Table 7. Comparison of financial costs in waste management

Relative cost of	Hong Kong (HK$/per capita GDP/month)	Australia (A$/per capita GDP/month)	Guangzhou (RMB/per capita GDP/month)
Landfilling each tonne of waste[#]	0.65%	0.55–0.68%	0.96%
Capital costs of landfill construction	0.26%	0.24–0.29%	0.97%
Collecting each tonne of domestic waste*	3.3%	n.a.	3.6%

* Includes basic collection cost and the cost of refuse transfer; #excludes the land cost; data for Hong Kong derived from PELB, 1998 and EPD 1999c; data for GZ derived from Guangzhou Construction Committee and Guangzhou Environmental Health Bureau, 1999; data for Australia derived from Australian Bureau of Statistics, 1999 and Xu et al., 1999. All GDP data are in current prices.

Guangzhou.[30] This suggests that people in Hong Kong are paying less for waste management services than their counterparts in Guangzhou and probably Australia as well. In sum, a strict comparison of the relative costs of landfilling shows that HK performs better than GZ. The cost performance of GZ would be even worse if one were to take into consideration its poor performance of MSW.

Facilitating environmental citizenship in waste management

There is a growing tendency for wider adoption of a participatory approach in the management of solid waste in GZ. Recently a number of initiatives have been taken to augment public involvement and to identify popular demands and preferences. First of all, a series of pilot studies have been undertaken on source separation of domestic waste where household cooperation is actively sought. Secondly, opinion surveys have been conducted to collect popular views on waste separation. Thirdly, people in GZ are now provided with more complaint channels to redress their grievances against improper waste management. However, under a top-down mode of public participation, the effectiveness of these participatory arrangements is still doubtful. For example, public opinion surveys are mostly used for justifying the municipal government's waste policies as they are conducted after the policy decision is made. The use of complaint channels is even more limited as past figures have shown that GZ people are not very vocal on waste management issues. According to the EPB, complaints on solid waste pollution ranged from a low of 7 cases to a high of 33 cases per year in the period between 1987 and 1997.[31] However, this lack of complaints does not mean that people are satisfied with the agencies' performance in solid waste management as most local citizens have expressed the need for an improvement in environmental hygiene in Guangzhou (Guangzhou Environmental Health Institute and Guangzhou Public Opinion Research Centre, 1998, p. 2). On the whole, public participation in waste management is hardly adequate and effective when one considers that public consultation is still not a regular feature in the formulation of waste management policies and measures, information on service performance is not accessible, and transparency in policy making is low.

The situation is not much better in HK despite the availability of education and information channels in waste management at both policy and operation levels. The adoption of a participatory approach in the Hong Kong waste management system still falls short of facilitating environmental citizenship and attracting responsible behaviour for two reasons: the policy process is still agency-dominated and there is a lack of active agency support in the operation of SSPs.

Throughout the public consultation process, the public agencies have the discretion to decide the format and content of consultation. In extreme cases, public consultation can be abridged or conducted in a restricted way for the sake of administrative expedience. Both the District Boards and the ACE are mostly advisory setups and the former two municipal councils did not exercise direct jurisdiction over the two urban services departments.

Another major drawback is that the impact of public opinion on waste policy in most cases is inconsequential. For example, both the warnings of the landfill crisis and the suggestions of a shift of emphasis from disposal to waste reduction and recycling were ignored in the consultation of the Draft Waste Disposal Plan in 1989 and the strong public demand on higher order waste management needs has not been given due consideration. In addition, the transparency of the regulatory process is far from adequate and within a non-democratic setting stressing the value of an executive-led mentality, the receptivity of public agencies to popular inputs is limited, thus undermining public accountability of the HK waste management system.

Citizen participation at the operation level of waste management is even less encouraging. Despite strong demand, there is little progress in involving waste generators in waste management, nor is there adequate support for voluntary source separation of MSW leading to disappointing recovery performance (see "Participation in waste management by the public," p. 199 and "Policy contents in waste management," p. 193).

The following is a tabular summary of the above evaluation.

Table 8. Tabular summary of the performance of GZ and HK in waste management

	GZ	HK
Environmental impacts	• both direct and indirect environmental impacts are widely noted	• direct impact is controlled • indirect impact is more serious
Implementation effectiveness	• weak implementation, reflected in i) preparations for future waste facilities; ii) regulatory effectiveness of low and higher order waste management measures	• weak implementation, reflected in i) meeting some lower order waste management; ii) poorly planned and enforced SSPs
Cost effectiveness	• characterized by heavy government subsidization * cost-effectiveness low in collection and disposal of waste	* characterized by heavy government subsidization * cost-effectiveness high in MSW disposal and collection
Environmental citizenship	* not a regular feature	* public consultation is a regular feature of the system * but the policy process is still agency-dominated * a lack of agency support in popular waste management programs

Institutional constraints for sustainable waste management in HK and GZ

From the above analysis, it appears that the solid waste management system in Hong Kong is not yet prepared to meet higher order waste management needs, although it has been able to achieve quite competently lower order needs in waste management. In GZ, the agencies involved are aware of the need and urgency to improve all aspects of waste management, but any real effect has to be seen. This and the concluding section will consider the institutional constraints for HK and GZ to become sustainable and will answer the questions set out at the beginning of this paper.

Insufficient public consciousness of sustainable waste management

Public cognition of good waste management in Guangzhou is mostly restricted to keeping the environment free of litter and odors. Visual cleanliness and sanitation are the primary concerns in MSW management. An understanding of the issue of sustainability remains limited. As indicated by a recent study on the acceptance of the New Environmental Paradigm (NEP) by Guangzhou citizens, it was found that the NEP score of local people was a low 2.93, in comparison with that of 3.03 for American citizens in 1978 and for Hong Kong people in 1998. What is more significant is that about 5% of the respondents were unfamiliar with most of the NEP issues or statements (Chung and Poon, 2000). Despite the presence of general support for waste recycling, apparently there is a lack of understanding of the importance of source reduction or of stopping the abuse of environmental resources as judged by the NEP score. In short, more aggressive steps are required to motivate the public to work forward to a sustainable mode of waste management in Guangzhou.

Although public cognition of waste management in Hong Kong has gone beyond keeping the environment free of litter and odors to verbally supporting higher order waste management, an understanding of the issue of sustainability remains limited and empirical evidence shows that the commitment is low. For example, a recent survey on the abuse of plastic bags has found that the public has yet to develop source reduction and recycling habits despite an awareness of the urgent need to minimize the use of plastic bags.[32] With only limited exposure to SSPs, general citizens are not well informed on the importance, significance or the proper techniques for source separation.

The consumer based source reduction potential in HK is quite restricted, as shown by the low level of environmental awareness. It was found that the average overall acceptance of the new environmental paradigm (NEP) of local people in 1998 was similar to that of American citizens in the late 1970s and even lower than that of mainland Chinese in the rural areas during the surveyed period. Source reduction seems to be an even more remote concept among the public in the absence of any organized socialization effort and support from the government. In addition, the 'NIMBY' (not in my backyard) attitude toward waste disposal and treatment facilities is still quite common among the public. In short, similar to the case in GZ, aggressive environmental education programs are required to motivate local people in HK to become sustainability supporters. The limited understanding of more advanced concepts of waste management in the two cities has restricted the institutional changes necessary for the advancement of waste management.

Lack of innovative and long-term visions among waste management officials

Government officials in Hong Kong also do not have the proper mind-sets for sustainable waste management. Despite forewarning from the society, the past reliance on high standard sanitary landfilling as the main form of disposal and treatment alternatives for solid waste has underscored the agencies' lack of innovation and long-term vision in waste management. Another example is the reluctance of the Government Supplies Department to choose recycled products in major paper purchases at a time when recycled paper products are similar in quality and prices to virgin paper (see Griffin, 1994; Tsang, 1993). Thus the conservative waste management authorities in Hong Kong have not been alerted to the landfill crisis nor have they put forward active measures for higher order waste management. Indeed waste management has been narrowly viewed as pollution control for sanitation purposes through direct government regulation.

Despite signs of a paradigm shift to a sustainable orientation with greater consideration accorded to the waste hierarchy, public agencies have been slow to give extensive support to SSPs and even slower to take active measures, such as the producers' responsibility requirement and waste disposal charges on domestic and other wastes.

Government officials in Guangzhou are also not prepared to practice sustainable waste management. The waste agencies have been struggling to improve their performance in the collection, transportation and disposal of waste. The recent plans to diversify waste management options by introducing SSPs, recycling of putrescible waste and launching safe disposal projects are well needed but measures for source reduction/waste avoidance and the reuse of sources in immediate forms are missing from the environmental protection plans for the next century. On the whole, it appears that policies to translate advanced concepts in waste management are still out of the two waste agencies' frame of reference.

The limited use of economic instruments and the lack of economic incentives

Scholars have criticized the waste management hierarchy from an economic viewpoint and have urged for an economic weighing process for making waste management decisions (Karl and Rann, 1999).[33] While a fair economic weighing process is indeed needed, one of the preconditions for it is to level the playing field for an evaluation of waste management alternatives.

In the cases of HK and GZ, waste disposal is heavily subsidized, while waste recycling is expected to be a self-financing activity. More innovative forms of waste charging are potentially beneficial to waste reduction as illustrated by various overseas studies on waste charging and recycling performance.[34] However, despite being one of the freest economies in the world, the use of economic instruments to reduce waste generation/disposal has been limited in HK.

However, economic instruments that can be used to remove the economic imbalance are very limitedly applied in both cities. In HK, this is illustrated by the lack of economic incentives to reduce waste in the charging mechanism of waste collection and disposal. Currently, except in some exceptional cases, domestic waste collection, refuse transfer and landfilling are provided free of charge by the government. Industrial and commercial establishments as well as domestic households are required to bear the costs of collection in the form of a fixed lump sum. This cost arrangement makes them insensitive to waste generation. Moreover, the current institutional setting in Hong Kong is not favourable to a truly "polluters-pay" form of waste charging. On the one hand, the Legislative Council has been very critical of introducing new charges for public services during a time of economic uncertainty. On the other hand, the private sector has strongly opposed taking up the full cost of their pollution-generating behaviour. As a result, it was decided that the domestic sector will not be charged for waste disposal services (Waste Reduction Committee, 2000).

Other than the institutional barrier expected for this type of charging scheme, the charging level and format are also unable to prevent waste generators from avoiding waste production. The disposal charge figure used for non-domestic waste in the charging proposal under public consultation is HK$125/tonne which is well below the actual cost of waste disposal and does not consider landfilling replacement costs, environmental costs and the cost of using immediate waste facilities.[35] Thus, the waste disposal charge based on this formulation, even if approved for implementation, will not be able to generate any waste diversion or reduction incentives for most commercial and industrial waste producers.

There is a lack of economic incentives for the GZ district EHB and its personnel to control the increase of solid waste under existing input-oriented funding arrangements. On the one hand, the size of the district EHB's budget is linked to the quantity of solid waste that needs to be handled. On the other hand, the wages of individual workers in the collection and transportation of waste are decided by their workload, which means that they will receive higher remuneration if more waste is generated. Thus, the district EHB has not been keen to promote source reduction, even though the generation of excessive solid waste goes against the municipal government's position of pursuing sustainable waste management. This will not be a problem if the efficiency of supervision is high. Yet, at the operational level, monitoring is carried out by supervisory parties who can also gain from high waste production. Thus, under such circumstances, a performance-based pay is more in order to internalize the incentives for the EHB and its local agencies to actively control waste generation.

The existing charging systems for MSW in GZ have also failed to internalize incentive structures for seeking active cooperation from domestic households and individual business enterprises for waste reduction and recycling. The flat rate charge for household waste collection cannot offer adequate incentives for waste control. Similarly, the EPB is unaware of the need to attend to the economics of waste management. Its choice to use a ban instead of a product tax to eliminate non-biodegradable food containers is a case in point (see "Regulatory processes of waste management," p. 197).

In sum, the use of economic instruments is limited in both GZ and HK. Despite a wide range of options, which include variable rate waste disposal or collection charges, material charges, product taxes, deposit refunds, and recycling credits that can be used for waste control, the use of market instruments is limited to the waste collection/disposal charge and it is instituted in a form that fails to generate reduction incentives for waste generators. In fact, neither HK nor GZ has explored the full potential of applying economic instruments to attain more effective waste management.

The lack of active support for SSPs and recycling activities

In HK, support for SSPs and recycling activities from the government to both local communities and the private sector has been inadequate. The government has been passive in the promotion of source reduction and is reluctant to provide greater assistance to the growth and development of the local recycling industry under the traditional non-intervention approach of governance.

Although the number of SSPs has rapidly increased under government encouragement since the announcement of the WRFP, most have been voluntarily organized without active support or involvement from public agencies. As a result, source separation and recycling initiatives have met with little success due to weak program publicity, poor logistics of recyclables collection, non-sustaining financing and infrastructural support from the private sector — a result of the unlevelled playing field between waste disposal and recycling.

Despite the announced measure to assist the waste recovery industry, there is a lack of resource commitment on the part of the government to implement the scheme effectively. Most notably, the land allocation policy case demonstrates weak support for recycling within the government (see "Regulatory processes of waste management," p. 197). In addition, during the recent crisis in the paper recovery sector, the government has openly rejected the strong calls for the provision of economic assistance to the recycling industry (*Oriental Daily*, 1998).

Although waste recovery is also being discriminated against in GZ, the situation is less serious. Traditionally, the Chinese government has viewed secondary materials as a valuable resource. The Chinese government is directly involved in the trade and has also granted some tax incentives to the waste recovery sector (see Chung and Poon, 1998b). As such, the waste recovery sector is still able to thrive on the mainland and GZ citizens until now have been able to maintain the household waste recovery habit.

Concerning recent government initiatives to SSPs, a more flexible scheme is in trial in GZ, in that both door to door collection and bring forms are available to the residents and residents are able to obtain redemption money from the recyclable collectors, the SMO cleaning crew.

Thus, government-run SSPs in GZ are essentially formalized private recovery activities. The advantage of these SSPs is not so much in increasing the recyclables recovered, as GZ citizens are already recovering materials quite frequently, but in replacing the informal and unregistered waste recovery teams with more disciplined and better regulated ones.

Conclusion

From the brief review of the waste management situations in developed and developing countries in the first section, lower order waste management needs seem to be tackled prior to higher order needs. And whenever higher order waste management alternatives are instituted in developing countries, they are initiated by external bodies.

The present analysis shows that improvements in both lower and higher order requirements are taking place for MSW more or less at the same time in GZ, with the initiatives arising from within the governing bodies themselves. The evolution of waste management of HK follows the typical path of first meeting lower order needs then meeting higher order needs. Part of the reason for the more rapid institutional change from lower to higher order waste management in GZ stems from the fact that the waste recovery tradition (for both pre- and post-consumer recycling) is being retained in GZ, while in HK, the traditional process of polluting first and then cleaning it up haunts the waste management regime, making change more long-winded. In HK, while the green awakening to address higher order needs in waste management eventually came from within the body of governance, it was too little and too late.

The landfill crisis is the single most important factor in driving the public sector in HK. In GZ, the most important factor is to diversify waste management and to seriously consider alternatives, such as recycling and waste to energy. Public demand, in both HK and GZ, is at best a secondary driving force, if it is taken seriously at all by the authorities under the agency-dominated decision-making processes in the two cities. Environmental citizenship is not a well-founded concept in either city; the public at large is not well informed about sustainable waste management, nor is it prepared to commit to it. More disappointing, waste authorities in both cities have failed to effectively promote sustainable waste management, though for different reasons.

Despite the generally lower waste management standards in GZ and very unsatisfactory performance in the collection, disposal, and reduction in MSW management, there is one aspect in which GZ is ahead of HK: the waste authorities are able to institute higher order waste management measures, namely institutionalizing SSPs, more proactively and faster than in HK. For the GZ waste management authority, the establishment of government participated SSPs is an incremental institutional change while in HK it is more of a revolutionary step since consumer oriented and post-consumer recycling have already become things of the past. Moreover, the extent of governmental support required for SSPs in HK requires a leveling of the playing field between recycling and disposal activities. This again requires aggressive changes in the governance mentality of the concerned agencies. In contrast, in GZ the EHB can take advantage of the existing recycling habits of the residents and the economic motivations of the cleansing crews and temporarily suppress the adverse effects on SSPs arising from the unequal treatment between recycling and disposal activities.

Having said that, there is no complacency in either GZ or HK. Despite the green awakening of the waste management authorities in both cities, the commitments towards the waste management hierarchy are far from thorough and firm, whether it is in implementing and improving recycling or moving towards resource conservation and waste avoidance. To sum up, major constraints include a lack of extensive support from the society and the government, the reluctance to make use of economic incentives in the area of waste management, and the prevalence of a myopic mentality among the governing bodies. These constraints explain why a conflict between empirical phenomena and the waste management hierarchy is found in these two cities.

Notes

1. The waste management duty of the two councils has been taken up by the Food and Environmental Hygiene Department since early 2000.
2. Personal communication with P.C. Liang, on 28 February 2000, Guangzhou Environmental Health Bureau, Guangzhou.
3. The wage index of GZ in 1995 was 118.22 and in 1997 it was 123.95 (*Guangzhou Yearbook 1998*, 1998, p. 412; *Statistical Yearbook of Guangzhou 1997*, 1997, p. 370).
4. Judging from the way the recycling rates are presented in the document and previous communication with the Guangzhou shi feipin guanli bangongshi

(Recyclable Management Office) (see Chung and Poon, 1998b); however these are no more than guestimations.

5. A survey conducted by the Guangzhou Environmental Health Institute indicated that 77% of the respondents had the habit of recovering household recyclables (Guangzhou Environmental Health Institute and Guangzhou Public Opinion Research Centre, 1998). Another survey indicated that 85% of the population has such a habit (see Chung and Poon, 1999).

6. Chung (1996) found that 77% of the population in Hong Kong supports source separation of household waste while support from housewives is considerably less, at 59%. Concerning other waste prevention/reduction measures, such as redeeming bottles and imposing a product charge on plastic carrier bags, public support tended to be found among the more educated group. In another survey, Chung and Poon (2000) found that despite verbal support, only 18% of the respondents were recovering domestic recyclables materials at the time of the survey. This indicates a lack of continuity and commitment to recycling among the public at large.

7. By May, 1999, door-to-door collection of bagged waste services were offered in 92% of the households in GZ (Guangzhou Construction Committee and Guangzhou Environmental Health Bureau, 1999).

8. Personal communication with Y. H. Lin, on 23 March, 2000, YueXiu District Environmental Health Bureau and L.X. Liang on 3 April 2000, Tianhe Environmental Health Bureau. See also Note 2.

9. Safe disposal rates at the Da Tian Shan Landfill and the Li Keng Landfill are reported at 45% and 80% respectively (Guangzhou Construction Committee and Guangzhou Environmental Health Bureau, 1999, p. 341). Taking into account the disposal rate of the two landfills, the authors have derived the overall safe disposal rate of 63.4%.

10. Another initiative for waste recycling is giving concessions to private pyrolysis cum pelletised organic fertiliser plants, but this is only at an early stage of development.

11. See Note 2.

12. The recycling targets set in the Environmental Protection Plan for Guangzhou 1996–2010 are even lower than the guestimated waste recovery rate. While the overall waste recovery rate is guestimated at 40% and over 50% for metals, waste paper and plastic scrap, the targets set out in the plan for waste recovery is 20% by 2000. For detail, see Lo and Chung, 2000.

13. Those using the RTS as the disposal point of their waste are required to pay a charge equivalent to 25% of the recurrent cost of the RTS (PLEB, 1998). RTS private waste users represent a small proportion of total industrial and commercial waste generators. Thus, it would be fair to say that most of the waste disposal and transfer costs are paid by public money.

14. Chung and Poon (2000) found that 79% of the people in HK were not informed about the share of waste management in property management expenses.
15. The way the rest of the recyclables were handled was not documented in the report. However, according to the long-term research experience of the authors, the informal sector, (i.e., the building cleaners, haulers and scavengers) is responsible for capturing the other 70% of the recyclables.
16. This problem is found mostly in the EHBs of old administrative districts as these generally have a large proportion of senior staff who are entitled to many benefits and above average salary levels.
17. Enacted in 1997, the outright ban is imposed on the use, manufacturing and sale of non-biodegradable food containers.
18. Personal communication with P.Z. He and Y.H. Zhang, officer-in-charge and staff member of Guangzhou Recyclable Management on 16 July 1997.
19. However, this is not to say that the municipal councils are opposed to waste recycling. In fact, the two municipal councils and the municipal services departments have sponsored or are directly involved in a number of waste recycling programs, though all of them are either ephemeral or of very limited influence. Domestic waste recovery programs held by the municipal councils date back to 1989 when they administered a one-month plastic bag collection scheme. Another door-to-door household recyclable collection scheme was initiated the next year and lasted for 5 months (see Chung, 1996). In 1998–99, the former Provisional Regional Council sponsored a number of PET beverage bottle collection schemes and began to include the collection of source separated recyclables in its contracts for refuse collection services for two districts in May and November 1998 (Provisional Regional Council, 1999). In the same period, the former Urban Services Department also run a waste paper collection scheme.
20. In this regard the first piece of land was leased in 1998.
21. The land allocation policy has met with criticism for not tackling the major problem of the waste recovery industry, and because the tenancy period is too short and the land lacks basic facilities (such as power, water supply and site foundation) (see Chung and Poon, 1998b).
22. Public support for household waste recycling was found to have increased from 77% in 1992–93 to 95% in 1998. The same survey also found that the public preferred door-to-door collection of recyclables to the bring system which requires the householder to take the recyclables to a recycling bin located outside their residential buildings (see Chung and Poon, 2000).
23. See Guangzhou ershiyi shiji yicheng lingdao xiaozu (1998, p. 29, Table 4-1). Waste collection coverage is graded as unacceptable in the self-evaluation. The other aspects mentioned are graded as acceptable.
24. They are equipped with scrubbing and vehicle exhaust extraction systems and

some even have large-scale turn-tables to facilitate in-and-out movement of refuse collection vehicles.

25. There is another item, "nuisance", under the category of environmental health in the Urban Services Department's statistics that attracts more complaints from the public. The complaints categorized under "nuisance" consist of complaints about water dripping and seepage within buildings and from air conditioners. These nuisances do not arise due to poor municipal services and therefore should not be regarded as complaints. In the Regional Council statistical report, the term "sanitary nuisance" is used to denote similar complaints.

26. It was reported that 99% of the refuse vehicles could go in and out of the landfills within 40 minutes of time (EPD, 1998a, p. 98).

27. Direct communication with the cleaning crew, 12 January 1999.

28. Total waste recovery includes recovery from the construction waste stream.

29. About 20% of the households are free riders of waste collection services in Guangzhou (Personal interview with H. Li of the Environmental Health Institute on 19 April 1999).

30. In Xu, et al. (1999), the authors remark that not all landfills in Australia conform to the national requirements in providing the basic pollution control measures. However, all three landfills in HK were designed with state-of-the-art landfill pollution control measures.

31. Data supplied by the staff of the Guangzhou Research Institute of Environmental Protection.

32. In a popular survey conducted in May, it was found that although almost 70% of the people thought that it was not necessary to use plastic bags to package newspapers, 42% of the people still use plastic bags for newspaper procurement. The same survey also reported that over 65% of the respondents have not developed the habit of bringing their own shopping bags (Ming Pao, 2000).

33. Karl and Rann (1999) further cited Hecht and Werbeck (1998) in support.

34. Folz (1998) found that American cities using the "pay as you throw" type of waste charging method were able to sustain higher recycling participation rates than cities not using such a method.

35. At 1997 prices, the landfill cost was reported at HK$110/tonne (PELB, 1998). Recently, a newspaper report cited the cost as HK$125/tonne without specifying the time frame of the cost estimation (see Chan, 2000).

References

Agapitidis, I. and Frantzis, I. "A possible strategy for municipal solid waste management in Greece." *Waste Management and Research*, 16.3 (1998), pp. 244–252.

Australian Bureau of Statistics. *1999 Yearbook of Australia*, edited by W. McLennan. Canberra, 1999.

Blight, G.E. and Mbande, C.M. "Some problems of waste management in developing countries." *Journal of Solid Waste Technology and Management*, 23.1 (1996), pp. 19–27.

Chan, Q. "Recycling could earn cash reward." *South China Morning Post*, 2 March 2000, p. 1.

Centre of Environmental Technology. *Household Waste Surveys at Selected Housing Estates: Revised Final Report, Volume B*, FP 98-013. Hong Kong, April 1999.

Chung, S.S. *Policy and Economic Considerations on Waste Minimization and Recycling in Hong Kong*, Ph.D. Thesis. Hong Kong: Hong Kong Polytechnic University, June 1996.

Chung, S.S. and Poon, C.S. "Characteristics of waste and recyclables of Guangzhou and waste characterisation study." Working paper.

Chung, S.S. and Poon, C.S. "A comparison of waste management in Guangzhou and Hong Kong." *Resources, Conservation and Recycling*, 22.3–4 (1998a), pp. 203–216.

Chung, S.S. and Poon, C.S. "Recovery systems in Guangzhou and Hong Kong." *Resources, Conservation and Recycling*, 23.1–2 (1998b), pp. 29–45.

Chung S.S. and Poon, C.S. "The attitudes of Guangzhou citizens on waste reduction and environmental issues." *Resources, Conservation and Recycling*, 25.1 (1999), pp. 35–59.

Chung, S.S. and Poon, C.S. "A comparison of waste reduction practices and New Environmental Paradigm in four southern Chinese areas." *Environmental Management*, 26.2 (2000), pp. 195–206.

Conservancy Association. *Information Bulletin on Waste Reduction*. Hong Kong, June 1989.

Danteravanich, S. and Siriwong, C. "Solid waste management in Southern Thailand." *Journal of Solid Waste Technology and Management*, 25.1 (1998), pp. 21–26.

Diaz, L.F. "Alternative program offers insights on peri-urban waste management." *Worldwide Waste Management*, September 1999, pp. 19–21.

Environmental Protection Agency. *Environmental Protection in Hong Kong 1982–83*. Hong Kong: Hong Kong Government, 1983.

EPD. *Monitoring of Municipal Solid Waste for 1989–1990*. Hong Kong. Hong Kong Government, 1991.

EPD. *Monitoring of Municipal Solid Waste for 1991–1992*. Hong Kong: Hong Kong Government, 1993.

EPD. *Monitoring of Municipal Solid Waste for 1993 and 1994*. Hong Kong: Hong Kong Government, 1995.

EPD. *Monitoring of Solid Waste in Hong Kong 1995.* Hong Kong: Hong Kong Government, 1997a.

EPD. *Monitoring of Solid Waste in Hong Kong 1996.* Hong Kong: Government Printer, November 1997b.

EPD. *Monitoring of Solid Waste in Hong Kong 1997.* Hong Kong: Government Printer, January 1999a.

EPD. "1998 waste statistics update." (www.info.gov.hk/epd/E/pub/sw-rep/98update). 1999b.

EPD. Personal communication with the Waste Facilities Business Unit of the Environmental Protection Department, Hong Kong, 10 March, 1999c.

EPD. Personal communication with the Waste Facilities Business Unit, Facilities Management Group on 25 May, 2000, Hong Kong.

Faming, Y. and Wu, X.B. "Sustainable development in China's paper-making industry." *UNEP Industry and Environment,* Jan–June 1997, pp. 75–78.

FEHD (Food and Environmental Hygiene Department). "Pleasant Environment — Cleansing Services." (www.info.gov.hk/fehd/pleasant_environment/cleansing/clean1.htm). Hong Kong Government, 2000.

Folz, D. "What explains program success?" *Resource Recycling,* September 1998, p. 29.

Griffin, K. "Hong Kong poser on paper recycling." *South China Morning Post,* 10 January 1994, p. 6.

Guangzhou Construction Committee and Guangzhou Environmental Health Bureau. *Guangzhou shi huanjing weisheng zhong de guihua (Environmental Hygiene Planning and Design for the City of Guangzhou).* Beijing, 1999.

Guangzhou Environmental Health Institute. *Planning to Abate Pollution from Solid Waste in Guangzhou, 1995–2010.* Guangzhou, 1996a.

Guangzhou Environmental Health Institute and Guangzhou Public Opinion Research Centre. *Shenghuo laji fenglei shoujing minyi diaocha baogao (A Report on the Public Opinion Survey on Municipal Solid Waste Separation).* Guangzhou, November 1998.

Guangzhou ershiyi shiji yicheng lingdao xiaozu (Steering Group of Guangzhou's Agenda 21). *Guangzhou ershiyi shiji yicheng (Guangzhou's Agenda 21).* Guangzhou: Guangdong keji chubanshe, 1998.

Guangzhou ribao (Guangzhou Daily). 12 January 1999.

Guangzhou shi renmin zhengfu *(The Municipal Government of Guangzhou). Guangzhou shi huanjing baohu guihua (1996–2010) (The Environmental Protection Plan of Guangzhou [1996–2001]).* Guangzhou, 1996.

Guangzhou Yearbook 1998. Guangzhou: Guangzhou Yearbook Press, 1998.

Guangzhou Yearbook 1999. Guangzhou: Guangzhou Yearbook Press, 1999.

Hawthorne, M. and Alabaster, T. "Citizen 2000: Development of a model of environmental citizenship." *Global Environmental Change,* 9 (1999), pp. 25–43.

Hecht, D., and Werbeck, N. "Europische Abfallpolitik." In *Kompendium der Europischen Wirtschaftspolitik*, edited by P. Klemmer and R. Waniek, pp. 219–317. Munich: Vahlen, 1998.

Hoberg, G. J. "Technology, political structure, and social regulation: A cross national analysis." *Comparative Politics*, April 1986, pp. 357–376.

Hong Kong Annual Report 1999: A Review of 1998. Hong Kong: Government Printer, 1999.

Jiang, Z.Q. "Lun qingjie shengchan de tuixing jizhi" (Examining the Implementation of Cleaner Production). *Zhongguo huanjing guanli* (Chinese Environmental Management), 1 (1996), pp. 31–34, 37.

Karl, H. and Rann, O. "Waste management in the European Union: National self-sufficiency and harmonization at the expense of economic efficiency?" *Environmental Management*, 23.2 (1999), pp. 145–154.

Lo, C.W.H. and Chung, S.S. "Solid waste management in Guangzhou: The stumbling block to a world class city." In *Guangdong in the Twenty-First Century: Stagnation or Second Take-Off?*, edited by Joseph Y.S. Cheng, pp. 401–438. Hong Kong: City University Press, 2000.

Lu, C.Y. "An investigation on the municipal waste pollution measures." *Huanjing yu weisheng* (*Environmental and Hygiene*), 2 (1997), pp. 28–33.

Ming Pao. "The environmental awareness of Hong Kong people is low." 8 May 2000, p. A5.

Nanfang ribao (*Southern Daily*). 7 October 1998, p. 1.

Oriental Daily. "Seventy percent of waste paper collectors stop collection from today onwards." 3 December 1998, p. A16.

PELB (Planning, Environment and Lands Branch). *White Paper: Pollution a Time to Act*. Hong Kong: Government Printer, 5 June 1989.

PELB. *Waste Reduction Framework Plan*. Hong Kong: Government Printer, November 1998.

Provisional Regional Council. *Regional Council Annual Report 1998–99*. Hong Kong, 1999.

Provisional Urban Services Department. *Statistical Report 1997/98*. Hong Kong, 1998.

Provisional Urban Services Department. Personal communication with P. Poon, 15 December, 1999a. Hong Kong.

Provisional Urban Services Department. *Statistical Report 1998/99*. Hong Kong, 1999b.

Regional Council. *Regional Council and Regional Services Department Quarterly Statistical Report 1998 July–September*. Hong Kong, 1998.

Regional Council. *Regional Council and Regional Services Department Quarterly Statistical Report 1999 January–March*. Hong Kong, 1999.

Regional Services Department. Personal communication with L. S. Ho, 7 December, 1999. Hong Kong.

Schall, J. *Does the Solid Waste Management Hierarchy Make Sense? A Technical, Economic and Environmental Justification for the Priority of Source Reduction and Recycling.* Working Paper Series #1, Program on Solid Waste Policy, School of Forestry and Environmental Studies, Yale University, October 1992.

Statistical Year Book for Guangzhou 1997. Beijing: China Statistics Press, 1997.

Szelinski, B. "The new waste avoidance and Waste Management Act (WMA)." *Resources, Conservation and Recycling,* 2 (1998), pp. 3–11.

Tsang, D. "Recycling looks good on paper." *South China Morning Post,* Letters to the editor, 30 November 1993.

United Nations Conference on Environment and Development. *Agenda 21,* Conches, 1992.

UNESCO. *Intergovernmental Conference on Environmental Education,* Tbilisi (1977): Final Report. Paris: UNESCO, 1978.

UNESCO. "Teaching global change through environmental education." Connect: *UNESCO-UNEP Environmental Education Newsletter,* 18 (1993), pp. 1–4.

Urban Environment Company. "Solid waste management in Hanoi, Vietnam." *Warmer Bulletin,* 44 (February 1995), pp. 2–4.

Urban Services Department. *Statistical Report 19967/97,* Hong Kong, 1997.

Wang, J.Q. "Zaozhi chulu zai hefang?" (Is There a Way Out for Paper Making?) *Zhongguo huanjing bao* (China Environmental News), 7 Dec. 1999, p. 2.

Waste Reduction Committee. *Waste As Resources,* Issue 2, September 1999. Hong Kong, 1999.

Waste Reduction Committee. *Annual Review — Waste Reduction Framework Plan.* ACE Paper 3/2000, January 2000. Hong Kong, 2000.

Xu, X.L., Rudolph, V. and Greenfield, P.F. "Australian urban landfills: Management and economics." *Waste Management and Research,* 17 (1999), pp. 171–180.

Yhdego, M. "Urban solid waste management in Tanzania: Issues, concepts and challenges." *Resources, Conservation and Recycling,* 14 (1995), pp. 1–10.

Zhao Y.Q. "Baise wuran nineng nata zengyang?" (White Pollution, How Can You Tackle It?) *Yangcheng wanbao* (Yangcheng Evening News), 15 August 1998, p. A7.

PART IV

Governance

Shenzhen–Hong Kong as One: Modes and Prospectus of Regional Governance in the Pearl River Delta

David K. Y. Chu, Jianfa Shen and Kwan-yiu Wong
Department of Geography, The Chinese University of Hong Kong

Introduction

The release of a survey report by the Hong Kong–China Relation Strategic Development Research Fund sheds new light on changing sentiments on future housing locations of Hong Kong citizens (*Singtao Daily*, 2000). In the past, the vast amount of land north of Lowu was considered suitable only for the retired and the few who consider Guangdong to be a desired place of residence after retirement. To facilitate such needs, the Hong Kong government has given approval to Hong Kong elderly citizens to collect their social assistance and comprehensive aid if they live in Guangdong and only frequent Hong Kong occasionally. Released on the same date was another survey conducted by the Democratic Alliance for the Betterment of Hong Kong. Among the 427 interviewees, the poll showed that 65% of the respondents wanted to stay in the HKSAR (Hong Kong Special Administrative Region) after retirement and almost 80% said they would be unable to adapt to the way of life if they were to retire on the mainland (*Singtao Daily*, 2000). One major reason is that 75% of the respondents worried that medical services were sub-standard on the mainland.

This research is funded by the Research Grant Council of Hong Kong, RGC Reference No. CUHK4017/98H (Geography).

The findings of the survey by the Hong Kong–China Relation Strategic Development Research Fund confirm that more than 15% of the respondents (a total of 1,121 people successfully interviewed over the telephone) will have an interest in moving across the border in the next ten years. That is to say, with respect to the total population in the HKSAR, more than one million local residents are interested in moving their homes to Shenzhen in the next decade. If the findings are valid, this indicates a great change and the impacts of these changes on the future of Hong Kong will be profound and will require immediate action.

Although the survey did not distinguish between those wanting a second mainland home and those wanting to live permanently in Shenzhen, the Fund Chairman, Cheng Yiu-tong, suggested that low-income earners want to live on the mainland where costs are lower and middle-income earners want to find holiday houses in cities close to Hong Kong. From the statistical breakdown of the respondents, those with incomes below HK$5,000 a month in Hong Kong have a greater tendency to leave Hong Kong for the mainland. There is not much difference among the other income brackets. It is most surprising that those aged over 61 and below 18 have the lowest propensity, while those from ages 19 to 60 have relatively similar responses, ranging from 16.5% to 18.6%. The general response among the working population that over 15% of them prefer to move to Shenzhen in the next 10 years confirms the necessity of taking this into account in the future planning of Hong Kong (*Singtao Daily*, 2000).

The rush to the north has become an issue of popular debate only recently with the results of the above surveys. However, the problems of border control and the need for cooperation between the two "authorities" on immigration have long been recognized (Chu *et al.*, 1999). The history of cooperation on this issue dates back to pre-1997. The problem of loosening border control, or deregulation, is a new issue in urban governance and a challenge to both sides of the border.

Policy areas requiring regional cooperation

Apart from the tendencies for massive resettlement of Hong Kong residents north of the border which have been briefly outlined above, the most urgent and hottest issue concerning border control is the dramatic rise in passenger traffic during extended holidays and weekends. For example, over 100 immigration counters on the Hong Kong side are needed to cater to the peak rush hour of over 20,000 entrants or exits per hour. Still, it cannot solve the occasional jamming and resultant confusion. On Good Friday of April, 2000, for example, there were thousands of north-bound Hong Kong residents stranded at Lowu and other railway stations for over 5 hours. On April 27, 2000, the Sunday before Easter Monday, over 10,000 were still at the gate in Shenzhen before the normal closing time of 11.30 p.m. With full alert, fortunately there was no confusion on Easter Monday, the last day of the extended weekend when 193,104 people returned from the mainland through Lowu. The corresponding figure for Lok Ma Chau was 30,259 (*SCMP*, 2000a). The deployment of extra officers from the head office to handle the holiday traffic and the extension of the hours by delaying the closing time, for example in the early morning of April 23, 2000, and during May Labour Day of this year (*Ta Kung Pao*, 2000), are evidence of the cooperation of the border control authorities on both sides.

Progress for further extending the opening hours of the border and procedures to simplify the customs check have reportedly been made and a consensus on this issue seems in sight, leaving some technical problems still to be solved. Evidently, pressure to keep the Lowu border checkpoint open round the clock has now decreased and technical difficulties raised by the KCRC (Kowloon Canton Railway Corporation) and the Hong Kong government have been understood. The cost and effectiveness of opening the Lowu checkpoint have been questioned by many as well, for example in an editorial of the *Hong Kong Economic Journal* (2000a). The Lok Ma Chau road crossing is now considered to be a better alternative. Since the Lok Ma Chau checkpoint has already been opened round the clock for empty trucks, it would be possible, with some additional resources, to keep it open for passenger traffic as well. Passenger coaches are by far more flexible and economical for the relatively low volume of passenger traffic at late night. Yet, with the running of overnight coach services through the Lok Ma Chau checkpoints, the flow of passengers will be facilitated and psychologically the border will become much less of a deterrent to the residents on both sides.

However, to cooperate on handling the problem of passenger traffic on extended weekends or on normal days is only the tip of the iceberg of the problems of cooperation that are facing Hong Kong and its counterparts to the north of its border. Over the last decade, in public speeches on various occasions, one can identify in the minds of officials and town planners

Table 1. Areas of common concern and cooperation

1. Counteracting illegal and criminal activities and immigration
2. Cross-border passenger and freight movement
3. Environmental protection and improvement
4. Development of high-tech research
5. Tourism
6. Commercial exhibitions
7. Finance
8. Cross-border residence
9. Culture and education

common areas of concern between the two governments in the north and south of the border. These areas are presented in Table 1.

The priorities, however, may not be the same. Table 1 more or less reflects the order from the Hong Kong perspective. A speech from the Shenzhen Planning Bureau, for example, listed the following priority areas for Shenzhen–Hong Kong cooperation — commercial exhibitions, transport infrastructure, finance, tourism and information industry (Zhang, 1996). This is understandable because the cities in the Zhujiang Delta (also known as the Pearl River Delta, PRD for short) are more pro-growth (see for example, Li, 1996) than Hong Kong and deeply involved in economic enterprises, while the government in Hong Kong is much less involved in business but is more involved in facilitating the running of business. On top of these differences, the customs and border control are in fact less of a concern to the local urban governments in the Pearl River Delta than a concern to the central government, while the Hong Kong government is involved at both levels.

In a certain sense, this is part of the legacy of the pre-1997 era and part of the problem of real and potential competition between the two sides. The difference in the two systems under the "One Country, Two Systems" format, however, will sustain the continuation of a self-centered individualistic tendency. Indeed, there is no simple formula or ready precedence to follow that would constitute satisfactory solutions of all these differences. However, one precondition is obvious, that is, the need for more imagination, from Hong Kong as well as from Shenzhen, and also from other parties concerned, if we hope to arrive at a mode of regional governance that is more conducive to cooperation.

Modes of cooperation between Hong Kong and the Pearl River Delta

With the possibility of being biased towards Hong Kong, the essence of the relationship between Hong Kong and its northern neighbours, especially those in the PRD, depends very much on the role that Hong Kong assumes in the future and its effectiveness in playing that role. As explained later, it is not necessary that the relations be cooperative; they could also be non-cooperative, competitive, coopetitive (competition and cooperation at the same time), or even worse, malignantly competitive and predatory.

There is indeed plenty of room for discussion and imagination on the possible relationship between Hong Kong and its neighbours to the north before one comes to some form of cooperation. Indeed, one can start by examining the extreme cases before one settles on the possible modes of the in-between cases. It should also be noted that between the opposites, a "third way" of a totally different nature may exist that is not commonly perceived.

At the risk of over-simplicity, six modes of relationship between Hong Kong and the cities north of its border, including Shenzhen, are raised here and discussed. However, these six are by no means exhaustive and conclusive regarding the future regional governance of the PRD.

(a) Individualistic development with little or minimal coordination

Territorial-wise Hong Kong and the PRD belong to the same deltaic system. Politically and administratively, however, the region is very much divided. It was even more so when two separate, and potentially confrontational, authorities governed the two entities.

Before 1997, Hong Kong was governed by the British colonial government with a view that Hong Kong was potentially under a threat from the north. This besieged mentality dictated colonial planning and administration such that urban and economic development necessarily gravitated towards the portion south of Boundary Street in Kowloon Peninsula and Hong Kong Island. The development of the New Territories was best avoided, as it was a leased territory for 99 years and due to return to Chinese rule in 1997. A new town programme was launched in 1972 when urban congestion was so acute that it contributed to the riots in 1967.

Pioneering in the provision of public housing, those resettled in the new towns were largely the disadvantaged poor and the New Territories basically remained a reserved area for residential and recreational purposes. The industrial and commercial hub was largely confined to the two sides of Victoria Harbour. Interaction between Hong Kong and the mainland was very limited, to say nothing of cooperation between the two local authorities. Trade exchanges were allowed, with the supply of fresh water and fresh agricultural produce as the two major items and sources of revenue of the mainland government. Before 1978, civilian travel and industrial investments were not encouraged and kept to a minimum, mostly for non-profit making purposes. Official maps published in Hong Kong before 1984 had no details of the border area. Shenzhen in a certain sense was a non-place geographically.

On the other hand, the capitalistic path of development in Hong Kong was criticized by the PRD government before 1978 and in geographical textbooks and official descriptions, Hong Kong was not regarded as a part of the region. To the Guangdong authorities, Hong Kong was a non-place with all the vices of capitalistic sins.

(b) Individualistic development with some coordination

After 1978 more and more Hong Kong industrial enterprises moved to the PRD, leaving only a skeleton in Hong Kong. To pursue such a go-it-alone strategy of development as outlined in the previous section had turned to be not at all possible. With the signing of the Sino-British Joint Declaration in 1984, the strategy became outdated and a new strategy of development was required. It was highly unlikely that after the return of Hong Kong to Chinese sovereignty in 1997 that Hong Kong would once again return to this strategy under the "One Country, Two Systems". For the more outward-oriented Guangdong urban economies, Hong Kong has been the source of their "foreign" investments and a shopping window. Hong Kong is too important to be ignored.

With the signing of the Sino-British Joint Declaration, Hong Kong proceeded to a new phase of development. It began its transition from a British colony to a Chinese Special Administrative Region (SAR hereafter). Yet the say of the Hong Kong government was severely confined to technical matters. Important policy issues that were to have profound impacts beyond 1997, not to say political issues, were expected to be handled through liaison groups between the central Chinese government and the British government. Under the Sino-British Liaison Group, a limited number of subcommittees were formed to deal with these matters. Hong Kong government officials were enrolled as part of the British negotiation team on these technical issues. A slightly different and later group was the Sino-British Infrastructure Co-ordination Committee (Dec. 1994) set up to oversee applications of transport and town planning projects that would link Hong Kong with neighbouring cities. The committee was then replaced immediately after the handover by the Hong Kong and Mainland Cross-Boundary Major Infrastructure Co-ordinating Committee (ICC) on October 16, 1997, although not many of the subcommittees under the Sino-British Liaison Group continued to exist.

The profile of Hong Kong officials was enhanced by the formation of the Hong Kong-Guangdong Environmental Liaison Group in July 1990. This is often cited as an example of cross-border cooperation. Indeed, this committee met frequently and some fruitful outcomes were concluded and implemented, such as the straightening of the channels of the Shenzhen River. This committee was perhaps the only provincial-level committee before 1997 that was not staffed by diplomatic and political officials from both sides. After 1997, the Hong Kong-Guangdong Co-operation Joint Conference (CJC) was formulated in March 1998 with the hope that discussions of mutual interest would be held twice a year in Hong Kong and Guangzhou alternatively. The duty of the CJC is to "keep the two places informed of each others' development and to co-ordinate their policies" (*SCMP*, 1998a). In the perspective of Lu Ruihua, the then governor of Guangdong, the emphasis of the CJC should be placed on " trade and economic cooperation including infrastructure and information industry … exchange in the areas of education, technology and professionals; checkpoint establishment and management including a smooth flow of passengers, vehicles and freight" (*SCMP*, 1998a). The first meeting of the CJC was held in March 1998 and the second was held in September 1998. The conference led to an agreement on the extension of the opening hours of the land crossing, efforts to foster tourism and also measures to protect the environment (*SCMP*, 1998b).

Despite these efforts, overall coordination has been unsatisfactory. Other than the announced cooperative projects in 1998, there were few breakthroughs after 1998 (*Hong Kong Economic Times*, 2000). The number of passengers on weekends increased while the opening hours were not extended until the unbearable congestion of April 2000 and the resultant outcries from the citizens. Still, when the issue of strategic development and territorial planning for the "long-term", *i.e.*, until the 2011 review, were raised, the relationship between Hong Kong and the PRD was regarded as important but the details of cooperation were ignored. The vision, so to speak, is self-centered, without a true dialogue with the counterpart cities and the necessity of coordination. In many ways, this vision is a very much parallel to the plan prepared by the Guangdong provincial government in 1996. According to the Pearl River Delta Urban System Plan (PRDUSP), Guangzhou was planned to be the centre of the PRD city system as if Hong Kong did not exist and future development corridors were planned radiating from Guangzhou in various directions (Planning Commission of Guangdong, 1996).

(c) A hegemonistic Hong Kong with the PRD cities as satellites

On the opposite end of little or limited coordination between the two entities, there appears a vision of a grand regional plan. But it is yet to be decided who is in command and which city will command the grand regional system. With the risk of being Hong Kong-biased, the following discussion assumes Hong Kong is chosen as the hegemonistic city with the full backing of the central government. The discussion is speculative in terms of if, and a big if, of course, Hong Kong could have its way without considering what the rest of the region feels should be done.

A regional plan is nothing new to China, but it is certainly a novelty to Hong Kong. As Ng pinpoints, overcoming the administrative barriers to achieve coordinated planning and development is very difficult in the PRD which has two special administrative regions (Hong Kong and Macao), two special economic zones (Shenzhen and Zhuhai), the provincial capital (Guangzhou, which used to be the centre of the region) and the PRD economic open zone. "While Hong Kong as a small city-state has little experience in regional planning, high politics will definitely introduce more uncertainties in any endeavor towards regional cooperation and planning" (Ng, 1995: 23).

However, Ng may be wrong that Hong Kong is special in the Chinese system. Indeed, Hong Kong is the special administrative region of China under One Country Two Systems. However, how special is it? The Basic Law covers much about the internal governance of Hong Kong, but Hong

Kong has no say about national defense and diplomacy. There is also little on the relation between Hong Kong and other Chinese cities and the role of Hong Kong in the region and regional governance. In a certain sense, the specialty of Hong Kong is contextually dependent on the status of China in the world, the progress of globalization and perhaps the economic cycles. The strength and continued contribution of Hong Kong towards the national goals of the People's Republic of China also count. The opposition of the provincial government and lower level city governments to the dominance of Hong Kong is not unimportant but secondary in comparison with the primary national objectives and global trends.

One may thus dare to assume, at an extreme, that Hong Kong is privileged in winning the total support from the central government. The provincial and lower level city and county (*xian*) governments of Guangdong would be directed to cooperate with Hong Kong under a grand regional plan tailored towards the objectives of making Hong Kong a showcase of excellence. With the blessing of the central government, Hong Kong becomes a local hegemony under which the future demands and planning objectives of all other PRD cities are subordinate. The PRD would become the backyard of Hong Kong such that Hong Kong is the centre of prosperity enjoying the benefits while the hinterland absorbs the costs and pays the price for the prosperity of Hong Kong. In a certain sense, that Hong Kong is exempt from defense expenditures and tax contributions to the central government is a privilege reflecting the special status of Hong Kong in the rank and file of Chinese cities.

One would argue this mode of regional planning and governance is unrealistic and the central government would not allow it to occur. It is however not the intention of this paper to argue whether this scenario is the only alternative. With reason and social justice in currency, this mode of regional governance would be condemned by those with a conscience (how many are there in this corner of the world?) under the core-periphery analytic: the core intentionally under-develops the periphery by institutional forces and by trade.

Understandably, this mode of regional governance is not favoured by many. For the sake of argument, a grand regional plan dictated by the needs of Hong Kong would work wonder in special situations and for the very short term. In case of contingencies, for example, when the entire southern China regional economy, including Hong Kong, *were* in the midst of a great depression, and limited by resources and in need of a big push from the largest city, Hong Kong would be destined to take up this role in order to drag the whole region out from such a horrid state. A grand regional plan and a "supply-led" market economy do not sound appealing today but could be acceptable in a totally different context.

(d) A coerced consensus for Hong Kong as the dragon head

Not all Guangdong entities would resist the dominance of Hong Kong in the PRD system. Regardless, some low level cities and *xians* are still dominated. Their choice is which upper level city to be dominated by. For example, the cities of Dongguan and Huizhou are subordinate to either Guangzhou, or Shenzhen, or Hong Kong/Shenzhen. Either their interests are shared by the most benevolent arrangement or they are victimized.

Before 1997, Hong Kong was not yet considered. Guangzhou, the provincial capital city, was the prime city. All regional development plans at that time automatically gravitated towards Guangzhou. In 1996, the regional plan for the PRD was formulated under this mind-set, with Guangzhou euphemistically termed the "Dragon Head" of the region. Shenzhen occupied a far less important position in the plan because of its quasi-provincial status, built basically by the central government and ministries and not subordinate to the provincial government. The plan, said to harmonize the differences of interest of various cities in the PRD, eliminated some duplicated projects. It was well-intentioned, but its impacts were questionable. Plans were difficult to implement in the 1990s even by the central ministries because their relative strength had declined. Lower level entities, like those in the PRD, commanded many resources unregulated by their superior authorities and in a sense they were only responsive to market forces and their own egos.

In a certain sense, the situation in the PRD should be rectified. The lower level cities should be under better guidance in their decision-making processes and governance. Guangzhou has failed to lead and other than being high up on the provincial administrative hierarchy, it has lost most of its capacity to set standards — in terms of productivity, environmental quality, high-tech development, financial markets, etc. Shenzhen in many areas and industries has overtaken Guangzhou as the leading city in Guangdong.

In the meantime neither Shenzhen nor Guangzhou could take over the leadership of development in the PRD, not to say Guangdong and even a grander vision of the Xijiang (West River) Valley (for example, see He, 1997). Like Shanghai, Hong Kong is at the estuary of the Xijiang River while the Xijiang valley stretches from Yunnan, through Guangxi to Guangdong. Not only the PRD, but southwest China along the Xijiang River needs a " Shanghai" to act as a dragon head to take care of the dragon body and the dragon tail. Hong Kong is geographically and economically, but perhaps less so politically, destined to do so.

(e) Hong Kong as a world city and co-optition among all entities in the system

Instead of a coerced consensus through power plays to become the dragon head of the region — be it the PRD, the province, or the Xijiang River Valley, Hong Kong could take up the leadership as the world city of the region. By actively the extending its role to the hinterland, like New York or London, Hong Kong would exercise its superiority in the realm of information collection, resources distribution and enhancement of quality decision-making in the region. It is not by plan or administration, but by the success of its enterprises and management techniques that it can become the *de facto* "dragon head".

Leadership of this kind is subject to competition. In order that the rest of the cities cooperate with Hong Kong, it is necessary to compete fiercely with all real and potential competitors. By raising the threshold of market entry and demonstrating market controlling power, Hong Kong could then become the pivot of the region with the ability and desire to regulate the operation and direction of development in the region.

The example of the construction of the Chek Lap Kok airport is close to this. By investing heavily in a superb and ultra modern airport with enormous capacity, the hub of airport traffic in the region is now very much secured. It is very difficult for other airpots to challenge Hong Kong's status in air transport. Zhuhai International Airport is therefore a better place to host the International Airplane Expo than to be a competitor to the Chek Lap Kok airport in Hong Kong. If Hong Kong could excel and overshadow the rest of the cities in the region, just like Chek Lap Kok in the realm of airport operations and management, through ruthless co-optition, Hong Kong, the world city, would easily find willing (or less than willing) cooperative urban entities north of the border. They would have to join Hong Kong and listen to the wishes and desires of Hong Kong because they would have no chance to beat Hong Kong. Another example is the Zhuhai/Macao-Hong Kong bridge. It will never be built until Hong Kong agrees, irrespective of how desirable the bridge is from the perspectives of Zhuhai and Macao.

Realistically, it would be very difficult for Hong Kong to maintain an absolute advantage over the rest of the cities in the PRD, especially over the long term. The world city and co-option mode of regional governance is academic wishful thinking. The superiority of Hong Kong is on the decline. The authors will return to this point in a later section.

(f) True partnership with a network of cities, including Hong Kong as one of the nodes

By relying on goodwill and sincerity for a commonwealth, there are relationships beyond cooperation and coordination. A good example is the relationship among the members of European Union although they are at a totally different scale — international rather than regional and sub-regional. The partners respect each other's differences but are willing to sacrifice their own egos and self-interest for the common good. There is a strong tendency of resource pooling and market sharing.

For a successful partnership, progress is normally slow and incremental. It does not start with a grand vision or a great plan. It starts with simple items of common concern in order to build consensus and mutual trust. A forum for exchanging different opinions and understanding at academic and civil society levels precedes official protocols.

One of the latest developments in business enterprises is the popularity of network firms and the emergence of industrial districts, like that in northern Italy or Silicon Valley. The versatility of firms allows the free flow of information and flexibility becomes a source of competitive advantage. The firms that survive in the 21st century must be learning firms. Geographically, a region that is to prosper in the 21st century needs to transform itself into a learning region. The PRD region and Hong Kong are no exception.

The "shop at the front with the factory at the rear" has signified the successful combination of the factor advantages of Hong Kong and the PRD over the last two decades. This model needs intensive coordination and command at the firm level. But it does not require much from the local governments. An environment of deregulation allowing for free movement of goods and personnel is sufficient. Differences in official standards and policies are taken for granted and firms can circumvent the red tape and differences through informal channels and unofficial means. Such harmonization of policies and unified standards were unheard of and unrealistic when the British government and the Chinese government confronted each other.

The Asian financial crises awakened the region to the fact that the prosperity before 1997 was built on a bubble. A foundation of solid growth should balance manufacturing and the service economy. High-tech production and high valued-added services will eventually replace the labour-intensive "shop at the front with the factory at the rear" model. With that in mind, the cyberport and Disneyland are two breakthroughs that Hong Kong has launched to rescue its faltering economy. The financial crises also hit hard the export sector of the PRD cities and the legacies of the neck-break growth attempts of the last twenty years were exposed to the public. The seriousness of pollution on the environment in Hong Kong and the PRD is alarming, while the weaknesses in governance on both sides of the border have been revealed. In the realm of public affairs, corruption among cadres in the PRD and other abnormalities in land deals and bank loans have received attention from the central government. The cases reported in local newspapers are just the tip of the iceberg. Although the financial system of Hong Kong survived, only isolated cases of poor corporate governance have been revealed. However, after the bursting of the bubble and under a new agenda for future development, the social support system is unsustainable. The education system, the health and hygiene system and the housing provision system of Hong Kong are all regarded as bad fits in the light of the challenges of the Asian financial crises and the new economy. In other words, both Hong Kong and the PRD cities fail badly on the fronts of economic sustainability, environmental sustainability and social sustainability upon which the cherished ideals of sustainable development are founded. Is it now time to turn over a new leaf and to move on to a new mode of regional and urban governance in order to fulfill the goals of sustainable development?

Shenzhen as a special case

In all ways, in considering the role of Hong Kong in the regional governance of the PRD, Shenzhen stands out as a special case on several grounds:

a. its adjacency to Hong Kong and its sharing of a land border with Hong Kong;
b. its status as the largest special economic zone, other than Hainan province;
c. its ever rising rank on the administrative hierarchy of mainland China;
d. its successful transformation from the old economy to the new economy; and
e. its relative short urban history and its migrant culture.

One must add that if Hong Kong people are to migrate northward, Shenzhen is the first and most likely the largest attraction of the spillover of Hong Kong residents in the near future. Its role is so special that no one can confidently predict what its future will be. In the authors' opinion, regarding Shenzhen as a satellite of Hong Kong is a serious mistake. This is the reason for singling it out for a special section in the paper and for including it in the title of the paper as well.

After the publication of "Bringing the Vision to Life: Hong Kong's Long-Term Development Needs and Goals" by the Commission on Strategic Development (2000) Hong Kong began preparing a 2030 town plan and territorial development strategy. One wonders how the 2030 town plan will proceed if we do not have the slightest idea about the future relationship between Hong Kong and other cities in the PRD. In the previous sections, we have suggested six possibilities. Are we going to proceed alone with the current form of individualistic preparation of a town plan from now to 2030 with but some limited cooperation and coordination with Shenzhen and the PRD cities? Are we going to respond to the demand for the round the clock opening of the border checkpoint with reluctance? Are we always to act reactively and not proactively? For example, only when the air pollution became so bad did we make up our mind to clean it up in five years' time (*SCMP*, 2000b) and we are finally aware of the need to send the Secretary for Environment and Food to Guangdong to discuss and to see if they can help.

Our northern neighbours are growing very fast and at rates so fast (currently about 10%) that it is sometimes really beyond the imagination of Hong Kong observers. The double-digit growth rate for the last three decades has transformed Shenzhen from a small frontier town to the sixth largest urban economy in mainland China. The projection is even more shocking that by 2005, its GDP per capita will be US$7,000 and in another ten years, i.e., 2015, its projected GDP per capita will be US$15,000 (*Hong Kong Economic Journal*, 2000b). Given Hong Kong will be growing at much slower rate for the next fifteen years, one can project that Shenzhen will catch up with Hong Kong in per capita output in less than 20 years. This projection can be substantiated by the fact that Shenzhen now has about 4 million young and vibrant people from all over China. Most importantly, they are highly educated; one estimates that 20% of all the Ph.D.s in China are found in Shenzhen (*Hong Kong Economic Journal*, 2000c). Indeed, it is reported that sooner or later Shenzhen will become the fifth directly administered municipality in the PRC after Beijing, Shanghai, Tianjin and Chongqing (*Hong Kong Economic Journal*, 2000b).

With a profit tax rate of 15% (Hong Kong's is 16.5%), many incentives to hi-tech industries, government encouragement and a plentiful supply of skilled engineers and technicians, and willing workers, Shenzhen is very attractive to Hong Kong entrepreneurs. In the past, because of the bureaucracy, the Shenzhen government had been criticized as inefficient and rigid. With the introduction of young, educated open-minded cadres into key civil servant positions, international practices have been widely adopted and the efficiency gap between Hong Kong and Shenzhen has been narrowing. In some aspects, for example, urban greening and tree planting along important roads, Shenzhen is superior to Hong Kong.

The price level is about one half that in Hong Kong today, partly because of lower wages and partly because of lower land costs. Rentals and house prices are about one-tenth of those in urban Hong Kong and one-third of those in the northern New Territories (Table 2). The construction of the underground railway system and the new city-centre at Futian will in four years' time bring forth a new and modern city with its own identity. So it is very wrong to conceptualize Shenzhen as a dormitory town or a huge shopping mall for Hong Kong residents.

The one million Hong Kong residents who wish to buy flats and settle in Shenzhen (cited in the introductory section of this paper) were mistaken as they did not take into account the changing price levels between Hong Kong and Shenzhen. They considered Shenzhen as being inexpensive and attractive, so that their earnings in Hong Kong could support their families comfortably in Shenzhen. The wish to commute daily across the border is not new. There are now over 1,300 students who go to Hong Kong for schooling and return daily to Shenzhen (personal communication) and there

Table 2. The narrowing gaps between Hong Kong and Shenzhen

	1995	2000	2005	2010
Price level below HK by price in Shenzhen	65%	50%	30%	0–15%
Average monthly wage level	946	1,500	2,230	4,000
(Non-skilled)	775	1,200	1,583	2,560
(Professional ceilings)	6,000	12,000	24,000	36,000
Monthly rental for a decent flat of 1000 sq.ft	n.a.	2,500–4,500	4,000–7,000	5,000–7,000
Selling price per sq. ft.	600	600	1,000	1,200

Source: Fieldwork and personal communication, adjusted with estimates by different sources.

are ten to seventy times more adult commuters every day. The extended opening hours of the checkpoints would indeed help this group of commuters. With the narrowing of the gaps in housing prices and wage levels due to the success of hi-tech industrialization and rising productivity in Shenzhen, the projected price differentials (see Table 2) will disappoint most of the million potential movers from Hong Kong to Shenzhen. Either they have to move now or in later years they will have to move farther north than Shenzhen or they will not be able to afford to move at all. The settlement of say one-third of the million will immediately drive up the housing prices and wage levels of Shenzhen and make it unaffordable to the relatively disadvantaged Hong Kong residents whose primary reason for migration is their relative inability to remain competitive in Hong Kong. Those who are not competitive in Hong Kong will find it even harder to compete with the Shenzhen residents.

Shenzhen therefore is a partner to Hong Kong, if not even a stronger partner. The relationship between Hong Kong and Shenzhen is symbiotic. A twin-city format in ten years' time and the prosperity of Hong Kong in the long run will be dependent on the real growth in Shenzhen and not vice versa. The border between Hong Kong and Shenzhen will persist perhaps even after 2047 but in twenty to thirty years' time, one should anticipate truly "free" flows of the population between Shenzhen and Hong Kong. It will be two-way with similar comfort and ease, unlike the asymmetrical order of today. A joint "urban council" of some sort will be formed and shared by the SAR and SEZ governments. The exact format of the composition of the joint "urban council" will not be discussed here and it is a bit premature to project urban governance of a twin-city several decades later. There is no shortage of cases and examples overseas, ranging from the Association of Bay Area Governments of San Francisco Bay Area (Boynton, 1976) to Metro Manila of the Philippines (Laquian, 1995). The Commission on Strategic Development would perhaps be a more suitable institution to deliberate the pros and cons of these institutions.

Nobody has the crystal ball of tomorrow, but there is a great likelihood that regional governance in the PRD will become a holy or unholy alliance of Shenzhen/Hong Kong versus the remaining cities and towns in the PRD and perhaps the Xijiang River Valley, not just Hong Kong with the rest. The status of Shenzhen from a quasi-provincial to a direct administered municipality will make it "independent" from Guangdong province because a directly administered municipality is of the same rank as a province and almost on a par with the Hong Kong SAR. The relation between Hong Kong and the PRD will be indirect because there will be no more land border between Guangdong and Hong Kong. Still if the twin-city is to prosper, the PRD and perhaps south and southwest China will be its hinterland. Whether the concept of a core-periphery or a network/partner format applies will depend on the mode of regional governance adopted by the twin-city authorities.

Concluding remarks

It is too far-fetched to outline and recommend which mode of regional governance Hong Kong and its neighbouring cities north of the border should adopt. The major objective of this paper is to point out that regional governance is a significant element to position Hong Kong on its way forward. The report "Bringing the Vision to Life" has rightly identified that the PRD region is integral to Hong Kong's long-term vision. *"The increasing economic strength of this region will provide the impetus for a significant part of Hong Kong's future growth as well as creating a highly prosperous city-region that would be far more competitive globally than each city would be individually"* (Commission on Strategic Development, 2000:12). Further research on how to strengthen and integrate Hong Kong with the rest of the PRD city-region is needed. There appears to be no simple and easy choice to arrive at a mode of city-region governance that is efficient, environmentally friendly, fair and acceptable to all, let alone enabling and empowering to the weak and disadvantaged members in the city-region.

References

Boynton J. "Governance of the city region." In *Growth and Change in the Future City Region,* edited by T. Hancock, pp. 129–173. London: Leonard Hill, 1976.

Commission on Strategic Development. *Bringing the Vision to Life: Hong Kong's Long-Term Development Needs and Goals.* Hong Kong: Hong Kong Government, 2000.

Chu D. K. Y., Shen J. and Wong K. Y. "Trans-border complications and regional governance: The case of Hong Kong and Guangdong." Paper presented at the *Conference of Cities in Transition,* organized by Hong Kong Institute of Planners, Nov. 8–9, 1999.

He Qirui (ed.). *A Study of Development of the Xijiang River Valley of Guangdong and Guangxi.* Guangzhou: Guangdong Jingji Chubanshe, 1997.

Hong Kong Economic Journal. May 3, 2000a.

Hong Kong Economic Journal. Feb. 21, 2000b.

Hong Kong Economic Journal. Feb.12, 2000c.

Hong Kong Economic Times. Apr.17, 2000.

Laquian A. "The governance of mega-urban regions." In *The Mega-Urban Regions of Southeast Asia,* edited by T. McGee and I. M. Robinson, pp. 215–243. Vancouver: University of British Columbia Press, 1995.

Li Zibin. "Talks on the Forum on Shenzhen-Hong Kong Cooperation" (mimeograph). Delivered at the *Forum on Shenzhen-Hong Kong Economic Cooperation,* Mar. 7, 1996 at Grand Stanford Harbour View Hotel, Hong Kong, organized by One Country, Two Systems Economic Research Institute Limited, Hong Kong.

Ng M. K. "Urban and regional planning." In *From Colony to SAR: Hong Kong's Challenges Ahead,* edited by Joseph Y. S. Cheng and Sonny S. H. Lo. Hong Kong: The Chinese University Press, 1995.

Planning Commission of Guangdong. *The Planning for Urban Agglomeration of the Pearl River Delta Economic Region — Coordination and Sustainable Development.* Beijing: Zhongguo Jianzhu Gongye Chubanshe, 1996.

SCMP (South China Morning Post). Mar. 31, 1998a.

SCMP. Sept. 25, 1998b.

SCMP. Apr. 25, 2000a.

SCMP. May 10, 2000b.

Singtao Daily. April 10, 2000 (www.singtao.com).

Ta Kung Pao. Apr. 28, 2000.

Zhang Su. "Preliminary thoughts about the cooperation of major industries of Shenzhen and Hong Kong" (mimeograph). Delivered in the *Forum on Shenzhen-Hong Kong Economic Cooperation,* Mar. 7, 1996 at Grand Stanford Harbour View Hotel, Hong Kong, organized by One Country, Two Systems Economic Research Institute Limited, Hong Kong.

Characteristics of Governance in China Based on the Practice in the Pearl River Delta

Xueqiang Xu and Xin Wang
Centre for Urban and Regional Studies, Zhongshan University, Guangzhou

Introduction

As a new word in China, there is no perfect translation of the word "governance" into Chinese. Colleagues in Hong Kong have translated it as "*guanzhi* (管治)". Though not satisfactory, there seems to be no better translation, so we choose to adopt "*guanzhi*" as the proper translation.

As far as we understand, the introduction of governance arises from the increasing top-down restrictions from centralized state power that have failed to fit into a rapidly changing economic, social and cultural environment. It is proposed that the key to handling an affair successfully lies in making full use of the resources through dialogue, coordination and cooperation among all interest groups instead of solely relying on government power and control. Governance, therefore, emphasizes coordination rather than domination. The theory of governance, therefore, emphasizes public participation and dialogue, as well as coordination and cooperation among different parties concerned. Viewed from this perspective, the theory has positive effects, especially on environmental, regional and urban governance. Nevertheless, if state power is totally ignored it may move to the other extreme and result in negative consequences.

Different countries and regions have different historical backgrounds, cultural traditions and environmental conditions. Therefore, governance should be applied in different forms and models. One single model is insufficient to suit all countries. A fatal mistake might be committed if the model of developed countries is used in developing countries without any modification.

There is a long history of centralized power in China. The educational level of the majority of citizens and their democracy consciousness are very low and weak. The system of democratic decision-making is far from perfect and the objective conditions for public participation in politics are not yet very mature. However, great progress has been made in the last twenty years, with the process of policy-making increasingly becoming procedural, democratic and scientific. This paper will consider this issue based on the experience in the Pearl River Delta and Guangzhou.

The framework of governance

As mentioned earlier, the application of governance theory should vary among countries as a result of differences in historical background, cultural tradition and environmental conditions. The Chinese model of governance can be summarized as "government domination, multi-party participation, democratic negotiation and scientific decision-making".

Government domination

Some concepts of governance emphasize horizontal management, which however might decrease the validity and feasibility for the government to pursue long-term policies on key issues. This may lead to the transfer of decision-making to special interest groups without public control. This kind of governance without politics may improve the implementation of decisions in some areas for the short term. For the long term, however, it cannot solve the crisis of governance. From the perspective of its operation, the various elements in a changing multi-level structure can hardly be coordinated automatically without the centralization of state power. It is equally hard to imagine that such automatic coordination could give rise to a well-prepared plan and achieve a result expected by all interest groups. Currently the governments in western countries play a primary role in providing public products and services, stabilizing the macro economy, regulating social distribution and maintaining market order. The planned economy had been predominant in China for quite a long time, with high governmental centralization in managing social and economic life. It has been proved by practice, however, that such a planned economy is no longer in accordance with the development of social productivity. The socialist market economy has been put into practice since the introduction of reform and opening in China. The government plays an indispensable role in regulating the macro economy, providing public products and services and supervising the market. Some key issues in social and economic areas, such as eliminating the differences between regions, protecting the environment and natural resources, building the system of social services and improving the quality of the population, have to be regulated and managed by the government from long-term and overall considerations due to their complexity and variety. Neither the NGOs (non-governmental organizations) nor the public can take a leading role because they can hardly avoid being attracted by their current and individual interests. The proposed government domination is not in conflict with governance theory but it is more favorable for the implementation of governance. Government domination does not mean that the government takes care of everything. Instead, scientific decisions are made with multi-party participation and democratic negotiations under the leadership of the government.

Multi-party participation

In recent years, with the expansion of TNCs (transnational corporations), the extension of international trade and progress in information and communication technology, top-down state control has proved to be inadequate in the rapidly changing economic, social and cultural environment. Against such a background, decisions can be better carried out through coordination of all interest groups including the government, NGOs, enterprises and the public so as to effectively solve the problems of environmental protection, regional development and urban administration. Multi-party participation and coordination are generally acknowledged at present. The Chinese government has always attached great importance to public opinion and has always followed the line of mass consultation. The government listens constantly to the voices of the masses regarding infrastructure building, environmental protection and so forth. The forms of public involvement vary, including expert consultation, questionnaire surveys, colloquia, on-site meetings and interviews etc. The degree of participation also varies, ranging from direct participation to consultation. Multi-party participation has been incorporated into certain concerned laws and regulations. For example, the fifteenth clause of the *Regulations on Environmental Protection and Management of Construction Projects*, which was promulgated in 1998, clearly prescribes that preparation of the report on environmental impacts should solicit opinions from local units and residents.

Democratic negotiation

Between the government and the public and between different groups, the same problem may be viewed differently because of different perspectives. Governance emphasizes dialogue, negotiation and communication among different interest groups. Different groups exchange information and resources, communicate with one another, make concessions and reach one final common understanding. Many methods are used in China at present, such as expert consultation committees, questionnaire surveys and colloquia. For example, on-the-spot investigations and interviews were conducted to solicit opinions on the levying of land for the underground rail project in Guangzhou. As for the ticket price for the underground service, the urban government launched a discussion on TV and in newspapers and then came up with a fair price by integrating opinions from all sides.

Scientific decision-making

In view of the complexity of the social and economic problems, an appropriate, scientific and manageable decision are only be made through multi-party participation by the government, NGOs, corporations and the public via democratic negotiations. Opinions from all parties should be taken into consideration to arrive at a decision which promises long-run economic, social and environmental benefits.

Case studies

Public participation in environmental management — site selection of the Datansha power plant

It was stated explicitly in some relevant rules in 1993 that public participation is an important part in the evaluation of the impact on the environment. With the development of society, people are paying increasing attention to environmental problems. Associations of the environment industry that are composed of enterprises and institutions engaged in the production of environmental scientific equipment, project design and construction, and resource protection have been established in many provinces and cities. The associations are closely related to the environment bureau but at the same time each holds an independent status as a corporation. There are many environmental protection organizations, formed spontaneously by universities, that are composed of experts, professors and engineers. They all play an important role in the building of environmental protection projects and in the promotion of environmental protection. The site selection of the Datansha power plant is a good example.

(1) In the early 1990s, the planning, city construction and environment protection bureaus of Guangzhou planned to invest 800 million yuan to build a power plant on Datansha island. The plant would cover $84,000m^2$ and dispose of 900 tons of garbage per day. All advanced techniques for burning and environmental protection were planned to be imported from foreign countries and the project was to be constructed by foreign firms.

(2) The environment research institute at Zhongshan University conducted an evaluation of the environmental impact and concluded that the plant would have few negative effects and that the techniques were feasible. The problems that might occur included garbage spills during transportation, poisonous matter being released in the process of burning and environmental accidents due to improper management.

(3) The local residents were strongly against the construction of the plant and responded to the leaders and department concerned through various channels.

(4) According to the rules of the state environment bureau, a public questionnaire survey was conducted before construction. But the results did not truly reflect public opinion because most respondents came from the government. So the survey was considered invalid.

(5) The environment research institute at Zhongshan University conducted another questionnaire survey in 1999; 350 questionnaires were distributed — 150 in Nanhai (10 to city officials, 15 to government departments related to the project, 15 to the district government, 10 to general residents and 100 to affected inhabitants); 200 in Guangzhou (10 to city officials, 20 to government departments related to the project, 25 to the district government, 10 to general residents and 145 to local inhabitants). Among the respondents 85 percent knew about the project and the location

of the plant; 86.3 percent knew little about the techniques; 92.7 percent did not approve of or opposed the construction of the plant.

(6) As a result, the location was changed. But most people did not know the techniques based on the results of the survey. The reason for opposition was selfishness. It is worth studying whether it would have been better to change the location or to reinforce management and take measures to solve the problems.

The role of experts in regional governance — construction of the Sixianjiao Watergate

(1) The Guangdong provincial party committee of the CCP (Chinese Communist Party) and the provincial government decided to establish a high-level group to strengthen the planning, organization and regulation of the Pearl River Delta (PRD) economic region. The director of the group is Zhang Gaoli who also is a member of the Standing Committee of the CCP and executive deputy governor of the province. The deputy directors of the group include the principal leaders of Guangzhou, Shenzhen and Zhuhai, and the director of planning of the province. Other members include the mayors of every city and the heads of the provincial Construction Commission and the Science Commission. Under the group there is an office and an expert group. The expert group in charge of detailed organizing work is set up in the Planning Commission. It includes 57 persons and is divided into smaller groups such as an environment group, infrastructure group, industrial development group, urbanization group and social development group. The members of the various groups come from universities, research institutions and industries. The job of the experts is to take part in research and consultation.

(2) The planning for the PRD economic region was carried out from April 1994 to October 1995. It was divided into the promotion and organization stage, the initial planning and coordination stage, the compilation and coordination stage of overall planning, and the discussion and revising stage of overall planning. The coordination continued throughout the process of planning, including coordination of all departments, all cities and counties, as well as different subjects and programs. The experts played an important role in the coordination. Direct dialogue between experts and leaders greatly facilitated coordination.

(3) The Sixianjiao Watergate project was listed as one of the 13 planned

projects of the PRD economic region. The watergate is located at the front of West River and North River. In the 1960s, the proposal was put forward to build a watergate at Sixianjiao in order to control the water volume of the North River and West River, because during flood seasons, the water of the West River overflows into the North River and increases its anti-flood pressure. In the winter, the water of the North River flows into the West River and affects the urban water supply. The proposal was suspended because of divergent views among the transportation, river navigation and hydropower departments. After experts and department officers investigated on the spot and consulted constantly, they came to a consensus view that the advantages of the construction of the watergate would outweigh the disadvantages. The deputy governor, Zhang Gaoli, held a meeting and decided to start the project and he asked for a formal feasibility study to be conducted by the Pearl River water committee.

The role of people's delegates and members of the political consultative committee — the removal and rebuilding of Junyi football field

(1) All levels of the conference of people's delegates are supreme power bodies. The people's delegates are elected by the people and broadly represent the people. Furthermore, the delegates reflect the common people's opinions and take part in the decision-making of important problems. All sections of delegates, especially those from the intellectuals and democratic parties can make their voices heard. When the conference is in recess, a "through train" is practised such that the delegates can express the people's will in letters to the highest officials of all sections of the government.

(2) As the supreme decision-making institution of Guangzhou urban planning management, the meeting of the Land-use Office is mandated by the urban government, headed by the vice-mayor, and consists of all related departments. The Planning Bureau has the responsibility to collect the necessary information and to coordinate implementation among these departments. The meeting is currently held once a month.

(3) The planned building of an inner ring road would cut across Junyi football field which is the exclusive standard field in Dongshan district and the place where many well-known players often play football. The residents living nearby and the related department of Dongshan district opposed the plan vehemently. Their opinions were sent to the urban government. In

July 1996 the government decided to start the construction of inner ring roads simultaneously with the rebuilding of the field with the spirit of "to have one is better than not to have one." The government of Dongshan district was to implement the project.

(4) The design of the Junyi field rebuilding was designed under coordination from the Urban Construction Commission. It included 2 floors of underground parking, a 20 meter wide inner ring up-road, and a 200 meter small field. At the same time it restricted the height of buildings near the road to less than 100 meters. The Research Institute of Traffic Planning proposed a traffic organization plan. The Plan Bureau checked and ratified the red line to the Physical Culture Bureau of Dongshan district. However, the Urban Center Office of transport projects took over the field because of difficulties during construction.

(5) In October 1998, the delegates voiced many opposing ideas regarding the rebuilding of the field. They suggested directly to the executive mayor that a field built over the inner ring road would be "queer looking". Their opinion was to rebuild the field in the Second Commerce Bureau reserve land that lies southwest of Donghu park. The land had been allocated to build a commercial and residential building by the Second Commerce Bureau and developers. The executive mayor in charge listened to the opinions of the delegates and agreed in the land-use meetings that a standard field be rebuilt there. The land would be levied by the Urban Road Extension Office and then the field would be transferred to the Dongshan Physical Culture Bureau after construction was completed. The commercial and residential building could find land in other locations. Therefore, the Plan Bureau transferred the land to the Urban Road Extension Office. The plan and design of the field was finalized after coordination between the Urban Road Extension Office and the Urban Physical Culture Bureau in January 2000.

Conclusion

The introduction of governance theory is in line with the social development of the times. With its theoretical significance and practical applicability, it is increasingly acknowledged by more and more countries and international organizations.

As a developing country, China is developing her economy in a way that is characteristic of China according to Deng Xiaoping's theory. Ever since the introduction of the open-door policy, the reform of the economic and political systems in China is more profoundly carried out, also with the gradual perfection of democracy, laws and regulations, in order to meet the demands of the social, economic and development environment in an information age. Predictably, more and more non-governmental organizations will appear in China. With not only more occasions for their participation, but also stronger effects of their participation, the desire will become stronger for different interest groups and individuals to participate in policy-making. The Chinese government is improving its way of working. Currently, the government emphasizes being public servants and it pays special attention to public opinion. Policy is determined only after careful investigation and research work. On the other hand, the general educational level is not high among Chinese citizens, and it has yet to be improved. Consequently, public involvement in the policy-making process may be oriented to individual interests at the expense of long-term interests and the overall situation. In view of such a dilemma, it is proposed that the guiding effect of the government should not be disregarded in favor of the role of non-governmental organizations and the public. The most scientific policy can only be made through dialogue between different interest groups, which, through constant negotiation, learn from one another's strong points and offset their own weaknesses to reach a final common understanding.

References

de Alcantaran, Cynthia Hewitt. "Uses and abuses of the concept of governance." *International Social Science Journal,* 50.155 (1998), pp. 105–113.

Guangdong Construction Commission. *Plan for the Pearl River Delta Economic Region.* Guangzhou: Guangdong Economic Press, 1995.

Pierre, Jon. "Models of urban governance: The institutional dimension of urban politics." *Urban Affairs Review,* 34.3 (1999), pp. 372–396.

State Environment Bureau. *Training Book for Environmental Impact Assessment in China.* Beijing: Chemical Industry Press, 2000.

Stoker, Gerry. "Governance as theory: Five propositions." *International Social Science Journal,* 50.155 (1998), pp. 17–28.

12

Globalization, Industrial Restructuring and Shifts: The New Industrial Zone in Shenzhen

Chaolin Gu, Guo Chen, Chaoyong Huang and Feng Zhen
Department of Urban and Resource Sciences, Nanjing University

Weizhong Pan
Branch Bureau of Longgang District, Bureau of Planning and Land, Shenshen,

Jianfa Shen, David K. Y. Chu and Kwan-yiu Wong
Department of Geography, The Chinese University of Hong Kong

Introduction

In recent years, economic globalization has been the driving force behind the restructuring and transition of world economic and industrial structures (Storper, 1997). As one of the most outward-oriented economic regions in mainland China, Shenzhen inevitably will be affected by such a process (Leung, 1996; Smart, 1995; Soulard, 1997; Weng, 1998; Yeung and Chu, 1998). For a long time, industrial development has been banned from the Eastern Industrial Group area[1] in Shenzhen, which is reserved for future use. On the basis of the construction of a high-tech park, this paper proposes launching a new advanced industrial zone emphasizing manufacturing production.

    The new industrial zone will be launched based on three main considerations. First, after the return of Hong Kong and Macao to Chinese sovereignty, the regional function of Shenzhen as a "border" city has

The paper is based on research supported by the National Natural Sciences Foundation of China, Project No. 49831003 and Hong Kong Research Grant Council, RGC reference no. CUHK4017/98H (Geography).

obviously been reduced. It thus needs to find a new base for economic growth. The Longgang industrial base and the Eastern Industrial Group are regarded as a base for modern advanced industrial development in the future. Second, it is proposed to change the present model of "Sanlai Yibu" (commissioned outward processing and compensation trade), dominated by medium and small-sized corporations, to the development of large industrial enterprises, gradually turning this area into a modern industrial entity. Last, in order to keep pace with globalization and the restructuring of the world's industries with advanced technologies, Shenzhen will be able to provide space for development in this direction. This paper attempts to discuss the potential for the development of this new advanced manufacturing agglomeration.

Opportunities and challenges of economic globalization for Shenzhen

Economic globalization and the return of Hong Kong and Macao to Chinese sovereignty have brought an opportunity and also a demanding challenge to the economic development of Shenzhen. First, consider the opportunity. With the return of Hong Kong and Macao, chances have increased for Shenzhen to acquire much more foreign investment, advanced technology and information, which will expedite the transition of local labor-intensive industries to capital-intensive and technology-intensive industries. Furthermore, economic globalization provides a good channel for the export of traditional labor-intensive products in Shenzhen, which are competitive due to low costs and prices. Finally, economic globalization will change the vertical division of labor among Shenzhen, Hong Kong and other advanced countries and regions. A new scheme for an international division of labor will be formed mainly in a horizontal direction with a complementary vertical direction.

At the same time, Shenzhen will face some critical challenges in the process of globalization. First, with knowledge and technology updating themselves faster than ever, increases in development costs in science and technology have become a worldwide tendency, thus making innovation, commerce and management centers highly concentrated in a few global cities. As one of these cities in East Asia, Hong Kong has been strengthening its regional function, especially after its return to Chinese sovereignty, while the regional function of Shenzhen as a "border city" has been obviously reduced. Furthermore, local technology and capability in R & D are not strong enough to support sustainable and stable development in Shenzhen. Second, as one feature of economic globalization is the shift of manufacturing industries from developed to developing countries, it also makes low-wage products increasingly less competitive in the world market. Third, at the level of industrial enterprises, technology alliances for TNCs (transnational corporations) and mergers of corporations have become dominant trends in the world economy. More and more enterprises are forging on the road toward international development. The current hierarchy of corporate structure in Shenzhen, which is mainly comprised of medium and small-sized corporations, is clearly a disadvantage to its competitive role in the 21st century. Fourth, some new rules of world trade have disadvantaged the developing countries and regions including China. For instance, the Intellectual Property Protocol approved by the WTO (World Trade Organization) in 1995 has increased the transfer costs of intellectual property enormously. Therefore, Shenzhen will have to spend much more for foreign technology and patents. For another example, the measures taken by the developed countries for environmental protection have been much more complex and critical. Foods, textiles, machinery and electronic products lower than standard will be strictly limited on the import lists of the western countries. Thus Shenzhen is forced to adapt higher standards for environmental protection in future industrial development. Finally, all the member countries in the WTO were required to cut or partly relinquish information technology taxes before 2000. This provides a convenient way for developed countries with high-tech to enter China's market, but it will weaken the traditional industries in Shenzhen.

Impact of global industrial restructuring and the shift to Shenzhen

As concepts like information society, post-industrial society and economic globalization emerged in succession, technology has become the most important factor among the traditional factors in production. Economic globalization is primarily realized in the restructuring and shift of world industries. Basically, industrial restructuring refers to a structural transition, characterized by the replacement and suppression of dominant industrial sectors rather than simply proportional changes in some sectors. Industrial shifts, in the narrow sense, refer to shifts from the economic advanced countries and regions to the less advanced countries and regions, with the spatial mobilization of production factors. In general terms, an industrial shift is not only a spatial concept, but also has implications in shifts of factors, such as capital, the labor force and technology among different sectors. We can say that the restructuring and shift of industries are dual-dimensional dynamic processes in time and space and also economic phenomena with diachronic and synchronous features. If a higher industrial structure is the aim of the industrial transition, the restructuring and shift of industries can be viewed as dynamic processes of the industrial transition, during which production factors flow over time and space. The optimization of the industrial structure is not only a matter of proportional change, but also a matter of coordination among the primary, secondary and tertiary industries.

In the process of economic globalization, new industries have emerged one after another. During the last 25 years, IT, R & D, medical science, education and environmental protection have surged at skyrocketing speeds with seemingly infinite potentials for future development. As the concentration of traditional industries in developing countries has been accelerated, NIEs (newly industrialized economies) are engaged in a shift from labor-intensive industries, which have lost part of their comparative advantages, to capital-intensive and technology-intensive advantages. A transition in industrial structure is thus worldwide. As a result of the rapid development of science and technology, especially the development and application of high-tech, some traditional industries are able to survive in higher structures through upgrading and restructuring. The transition in the local industrial structure follows a shift from a descending priority of the primary, secondary and tertiary industries, to the secondary, primary and tertiary industries, and finally the tertiary, secondary, and primary industries. From the perspective of dominant factors, it is a shift from labor-intensive industries to capital-intensive and technology-intensive industries. From the perspective of dominant industrial sectors, it is a shift from light industries and machinery manufacturing to heavy chemical industries and high-tech industries.

Since the 1980s, the transition in the world industrial structure has accelerated. Basically, it is a distinctive phenomenon with the emergence of high-tech industries dominated by information technology and a structural transition into knowledge-intensive and technology-intensive industries. It has several detailed features.

(1) The high-tech proportion in the structure of traditional manufacturing — From 1985 to 1995, members of OECD saw a doubling
in the proportion of the production and export of high-tech
industries in total manufacturing sectors. Computer, electronics
and aeronautics industries expanded most quickly.

(2) The flourishing of producer service sectors — Now the producer
service sectors have become the main buyers of high-tech facilities.
80% of the information technology products in the U.S. and 70%
of the computers in the U.K. are ordered for producer services.
Technology costs in industries such as aeronautics, telecommunications, commerce, medicine, finance and insurance account
for more than half of the total costs.

(3) The high proportion of intellectuals was in the world occupational
structure — Technology specialists, especially those in production
service sectors have risen proportionally.

The restructuring of world industries has an impact on the industrial
development of Shenzhen in two aspects. In the positive aspect, it is an
impending task for Shenzhen to join the mainstream. In order to keep pace
with the world restructuring of industries and to provide space for the coming
manufacturing industries of the TNCs, Shenzhen has to restructure its current
industrial structure of "Sanlai Yibu" or transfer the traditional industries
into outer areas. In the negative aspect, there are risks. If Shenzhen cannot
keep pace with globalization and lags behind in the process of world
restructuring and shifts, chances are that a long period of local economic
recession will persist.

In the current condition of the world shift of traditional industries, trends
have not changed much in the direction of a shift from developed economies
to developing economies and from high industrial grade areas to low
industrial grade ones. In East Asia and South Asia, there has long been a
five-grade system of industries, comprised of Japan in the first grade, the
four "dragons" of Asia (Singapore, Korea, Hong Kong and Taiwan) the
second, Thailand, Malaysia and Indonesia the third, mainland China's
coastal area, India, and Vietnam the fourth, and Mongolia, Laos and
Cambodia the fifth. In the 1960s, Japan was successful in making space for
industries shifted from the U.S. and Western Europe, gaining a 10-year
GDP growth at an average rate of 10.9% per year. In the 1970s, it was the
turn of the four "dragons" to take the torch. They reacted a record 9.3%
annual GDP growth. Another 10 years later, the three countries of ASEAN

took the industrial restructuring in Japan and the four "dragons" as a chance to attract labor-intensive industries. There followed a remarkable 7.6% average annual increase in GDP from 1986 to 1990. Clearly, the coastal area in China is an important participator in this "torch game". In the process of the global shift, China achieved a nearly 20-year high-speed economic growth. As a result, Japan is the first in East Asia to enter the post-industrial society, which is characterized by an industrial structure dominated by manufacturing, information, biology and other high-tech sectors. In the middle-late stage of industrialization, the four "dragons" now form an industrial structure dominated by capital-intensive and technology-intensive industries. The coastal areas in China and the three countries of ASEAN, still in the prime-middle stage of industrialization, share the character of an economy dominated by technology- and labor-intensive industries. In the other countries and regions in East and South Asia, industrialization has just started, and the cheap labor-intensive industries still dominate their economies.

Currently, the world industrial shift exhibits new characteristics. First, a transition from the wave-like direction down the industrial grade hierarchy, to a sustained multi-directional dispersion of technology has been noted. Second, a spatial shift from individual corporations of different sizes to a production network linked by TNCs can be observed. Currently, 44,000 TNCs and their subordinate corporations contribute to 1/3 of the world output value and 2/3 of the world trade volume. Their overseas investment now accounts for 70% of the world's total. With TNCs playing leading roles, the worldwide shift will continue to provide great opportunities for Shenzhen's adjustment in industrial structure.

The return of Hong Kong and Macao to Chinese sovereignty: the economic integration of Shenzhen, Hong Kong and Macao

After the return of Hong Kong and Macao to the motherland in 1997 and 1999 respectively, Shenzhen has experienced a change from a spontaneous economic cooperation with the two regions to a multi-level cooperation regulated by the government and market. A platform has been formed for future development and economic integration. But currently, cooperation among the three regions still remains at the so-called "front shop and back factory" stage (Sit, 1998). Under such a vertical division of labor, Hong

Kong and Macao provide funds, markets, technical equipment, and the main materials and accessories, while Shenzhen provides a labor force, infrastructure and some of the materials and accessories, as the location of production.

Because the industrial gap among Shenzhen, Hong Kong and Macao has been narrowed, future space for their cooperation at a low level is rather small. More efforts must be made to expand the cooperation channel, raising the cooperation level and creating new cooperative forms. It is already known that Shenzhen has adopted a method of attracting foreign investment by "attracting business by business and then following up paradigms". Also there are introductions by friends or acquaintances. There is little government intervention in the process, which makes investors poorly informed about the investment environment and causes blindness in the decision-making of investors and to-be invested. But after the return of Hong Kong and Macao to Chinese sovereignty, things began to change. On the one hand, cooperation and coordination among them will build bridges for future development and reduce the blindness of investors and to-be invested. Further emphasis on government policies and regulations will form a multilevel cooperative system in all directions, changing the form of "front shop and back factory". On the other hand, consultation and coordination systems among the three governments in macro-economic regulation will build a whole market of funds, technology, labor force and production. When resources are rationally allocated, future development and economic integration will be facilitated.

Since the economic reform and opening in China, Hong Kong has been the main source of foreign investment in Shenzhen, accounting for more than 65% of the total foreign investment. Business people from Hong Kong have invested in nearly 10,000 foreign-funded corporations and 20,000 "Sanlai Yibu" corporations. So the restructuring and shift of industries in Hong Kong have far-reaching impacts on Shenzhen. For Hong Kong, the high cost of land and labor, traffic congestion and deteriorating environment have been the main constraints for future development. The external economy can only be improved through industrial restructuring and an industrial shift and synergy with Shenzhen in city functions. While for Shenzhen, proximity to the important world center of trade, commerce, information and shipping means a convenient channel and springboard to enter the world market. In the 21st century, Hong Kong will be highly concentrated with commerce, trade, information services, recreation and high-tech industries. Due to the high population-density and land scarcity in Hong Kong, a high-tech industry base will be extended to Shenzhen, promoting its industrial upgrading and transfer of traditional industries. This new kind of division of labor and cooperation will push Shenzhen and Hong Kong in the direction of development and economic integration.

Because of its comparatively small economic scale, Macao has an indirect and minor impact on Shenzhen's industrial development. Clearly Macao has expanded its economy from mainly gambling business to diversified businesses, including tourism, consulting services and real estate industries. It is possible that the future development of package tourism and joint ventures in high-tech are possible among Shenzhen, Hong Kong and Macao. Thus the dispersion and transition process of traditional industries in Shenzhen will be expedited, promoting the construction of the Eastern Industrial Group in Shenzhen.

Analysis of current industrial development in Shenzhen

Currently, industries in Shenzhen are "light, new and high-tech", with a high proportion in the electronics and light industries, quick development in new industries and a considerable scale in high-tech industries. In 1997, high-tech industries in Shenzhen had an output value of 47.4 billion RMB, which accounted for 35% of total industries. The basic character of the industrial structure can be summarized as "three more and three less". First, there are more "Sanlai Yibu" corporations and less TNCs. In Shenzhen 99% of the 80,000 total corporations are medium/small sized. Second, there are more labor-intensive corporations and less knowledge- or technology-intensive corporations. In 1997 there were only 124 registered high-tech corporations, accounting for 0.1% of the total. Third, there are more foreign-invested corporations than domestic-invested corporations. The former now contribute to more than 70% of the total industrial output value.

For a certain industrial sector, the location quotient[2] is an indicator of its comparative advantage and competitiveness, and also its status and role in the regional functioning. Using the standard of sector divisions set by the State Statistical Bureau, Shenzhen has 33 sectors of industry. The location quotients for these different sectors are calculated (see Table 1).

According to Table 1, the following points are clear. (1) When grouped by light industry and heavy industry, the location quotient of traditional light industries has declined. Some industries with a traditional advantage

Table 1. Location quotients for industrial sectors in Shenzhen

Sector	1995	1996	1997	Annual increase	
				95/96	96/97
By light & heavy industry					
Light industry	0.7736	0.749	0.7134	−0.0245	−0.0360
Heavy industry	1.3599	1.402	1.4601	0.0421	0.0580
By size of enterprises					
Large-sized	1.6468	1.7956	1.374	0.1488	−0.4220
Medium-sized	0.2826	0.2739	0.9775	−0.0087	0.7036
Small-sized	0.8833	0.7983	0.7997	−0.085	0.0014
By sector					
Non-metal mineral mining and processing	0.1844	0.1365	0.1495	−0.048	0.0130
Foods processing	1.4106	1.6547	1.3649	0.2441	−0.2900
Foods production	0.4061	0.3470	0.2746	−0.0591	−0.0720
Beverage manufacturing	0.5272	0.5867	0.5832	0.0595	−0.0030
Tobacco processing	0.5836	0.7866	0.8315	0.2030	0.0448
Textile industry	0.4057	0.3082	0.3003	−0.0975	−0.008
Garments and other fiber products	0.8367	0.7427	0.6045	−0.0939	−0.138
Leather, furs, down and other related products	0.3464	0.2499	0.2587	−0.0965	0.0088
Timber processing, bamboo, cane, palm fiber and straw products	0.8314	0.8264	0.9641	−0.005	0.1378
Furniture manufacturing	0.4062	0.4212	0.3316	0.0150	−0.09
Papermaking and paper products	0.6338	0.4822	0.3624	−0.1516	−0.12
Printing and record medium reproduction	0.8745	1.0564	0.9257	0.1819	−0.131
Stationery, educational and sports goods	0.8288	0.7215	0.5262	−0.1073	−0.195
Petroleum processing and coking products	0.0885	0.0795	0.0462	−0.009	−0.033
Raw chemical materials and chemical products	0.4237	0.34	0.3391	−0.0836	−1E-03
Medical and pharmaceutical products	1.3474	1.7491	1.8657	0.4017	0.1166
Chemical fibers manufacturing	0.2332	0.265	0.1914	0.0319	−0.074
Rubber products	0.2412	0.1416	0.0885	−0.0996	−0.053
Plastic products	0.4764	0.4593	0.3975	−0.0171	−0.062
Non-metal mineral products	0.5353	0.5176	0.4256	−0.0176	−0.092
Smelting and pressing of ferrous metals	0.5695	0.4561	0.4155	−0.1134	−0.041
Smelting and pressing of nonferrous metals	0.3334	0.409	0.4372	0.0756	0.0282
Metal products	0.6409	0.6806	0.6808	0.0397	0.0002
Ordinary machinery manufacturing	0.2492	0.1499	0.1217	−0.0993	−0.028
Special purpose equipment manufacturing	0.7312	1.1148	0.7322	0.3836	−0.383
Transporting equipment manufacturing	0.5116	0.4709	0.3218	−0.0407	−0.149
Electric equipment and machinery	0.0276	0.2069	0.2301	0.1793	0.0232
Electronic and telecommunication	2.5954	2.5122	2.5822	−0.0833	0.07
Instruments, meters, cultural and official machinery	0.5886	1.8279	1.5428	1.2394	−0.285
Other manufacturing	0.6043	0.3004	0.4075	−0.3038	0.1071
Electric power, steam and hot water production and supply	6.4314	6.6741	5.6888	0.2427	−0.985
Gas production and supply	0.1427	0.0692	0.0125	−0.0735	−0.057
Tap water production and supply	1.2688	1.1177	1.3058	−0.1511	0.1881

can remain unchanged with technology renovation. But others can move to outside areas, especially the Eastern Industrial Group Area. (2) When grouped by different sizes, large-sized corporations show a decrease in the location quotient during these years. This can be attributed to the decline of comparative advantage and the spatial transfer of some large corporations. (3) Among the 33 different sectors, the foods processing, medical and pharmaceutical products, electric equipment and machinery, instruments, meters, cultural and official machinery, electric power, steam and hot water production and supply, and tap water production and supply industries have location quotients higher than one. These industries surely have an important status and comparative advantage in the industrial structure. (4) There are sectors that have location quotients less than one and also exhibit a declining trend. These include the non-metal mineral mining and processing, food production, textile industry, garments and other fiber products, leather, furs, down and other related products, furniture manufacturing, papermaking and paper products, stationery, educational and sports goods, petroleum processing and cooking products, raw chemical materials and chemical products, rubber products, plastic products, non-metal mineral products, smelting and pressing of ferrous metals, ordinary machinery manufacturing and transporting equipment manufacturing industries. Due to factors like the rising cost of production and the transition of the city functions, these industries no longer enjoy competitive advantages and should be moved to outside areas.

Conclusion

Considering the accelerated process of economic globalization, Shenzhen should make efforts in technology renovation and upgrading of traditional industries, while at the same time putting more emphasis on high-tech industries and technology-intensive industries. Under economic globalization, some considerations for the future industrial development of Shenzhen are detailed as follows. (1) Based on the center of the high-tech park of Shenzhen, high-tech industries mainly in the fields of electronic information, new materials and bio-engineering should be highly promoted. (2) In order to provide space for future industrial development, we propose launching the Eastern Industrial Group as a new industrial zone with high international competitiveness. Progress is to be expedited in the technology renovation of traditional industries, in the R & D of advanced manufacturing technology, energy savings and clean production technology, and in advocating practical technology. According to world trends of industrial transition and current conditions in Shenzhen, we propose a combined strategy for the development of the Eastern Industrial Group, that is, to enhance the speed of growth of industries and to provide space for new industries, especially advanced manufacturing industries with a high proportion of technology and high industrial linkages. This is what Shenzhen should prepare in the process of global restructuring and shifts.

Notes

1. The Eastern Industrial Group is located in Longgang District of Shenzhen municipality, which includes the Longgang industrial base and the towns of Kengzi and Pingshan. The district has a total area of $169km^2$, of which $109km^2$ can be used, and $81\ km^2$ is planned for construction.
2. A location quotient higher than one indicates an advantage for the sector, while a quotient lower than one indicates a disadvantage for its development. An increasing number means a rising momentum for its development. The sector can be considered a "sunrise" industry. A decreasing number means a falling momentum. The sector can be considered a "sunset" industry.

$$\text{Location Quotient} = \frac{\text{Output value of a certain sector in Shenzhen / total output value of industries in Shenzhen}}{\text{Output value of a sector in Guangdong / total output value of industries in Guangdong}}$$

References

Leung, C. K. "Foreign manufacturing investment and regional industrial growth in Guangdong Province, China." *Environment and Planning A*, 28.3 (1996), pp. 513–536.

Sit, V. F. S. "Hong Kong's 'transferred' industrialization and industrial geography." *Asian Survey*, 38.9 (1998), pp. 880–904.

Smart, A. "Local capitalisms: Situated social support for capitalist production in China." *Occasional Paper No. 121*, Department of Geography. Hong Kong: The Chinese University of Hong Kong, 1995.

Soulard, F. *The Restructuring of Hong Kong Industries and the Urbanization of Zhujiang Delta 1979-1989*. Hong Kong: The Chinese University Press, 1997.

Storper, M. *The Regional World: Territorial Development in a Global Economy*. New York: Guilford Press, 1997.

Weng, Q. H. "Local impacts of the post-Mao development strategy: The case of the Zhujiang Delta, southern China." *International Journal of Urban and Regional Research*, 22.3 (1998), pp. 425–442.

Yeung, Y. M. and Chu, D. K. Y. (eds.). *Guangdong: Survey of a Province Undergoing Rapid Change.* Hong Kong: The Chinese University Press, 1998.

13

Study on the Innovation in the Administrative Organization and Management of the Metropolitan Areas in Mainland China, with Special Reference to the Pearl River Delta[1]

Junde Liu

Research Center of Administrative Division of China,
East China Normal University, Shanghai

Metropolitan areas play a very important role in the process of economic and social sustainable development and modernization in China. In a period of economic transition with the "jurisdictional economy" in operation, many problems in the present spatial organization and management system have appeared. To a certain extent, these problems constrain the further development of metropolitan areas in mainland China (Liu and Wang, 2000; Liu, 1999; 2000). In this paper, on the basis of an analysis of the administrative patterns of metropolitan areas on the mainland, the author points out the way for the reform of the spatial organization and administrative system in the country.

* The paper is part of the research results from a project funded by the National Natural Science Fund: <Study on the Feasible Scheme of the Pattern of Spatial Organization for the Control by Different Levels of the National Economy in China> (No.49771025).

The Development of Metropolitan Areas in Mainland China and the Main Types of Administrative Organization

The development of metropolitan areas in China

In China, there is no strict definition of a metropolitan area (MA) at present. Usually, it is regarded as a region where a large central city or two or three cities at the core are integrated with several small cities at the periphery. MAs have been growing rapidly due to the great changes in the economic and social structure since 1979. By the end of 1998, there were 668 municipalities in the entire country. Among them, there were 37 metropolises, each with a population of more than one million, and 48 large cities with a population of 0.5–1 million each. If we define an urban region with a non-agricultural population of more than one million as a MA, then there are more than a total of 30 MAs in China. In regions with a high density of metropolises, there have emerged megalopolises or metropolitan interlocking regions (MIRs).

There are four MIRs in China, all located on the east coast. Their basic conditions are presented in Table 1. From north to south, these MIR's are as follows:

Table 1   Key indicators of four MIRs in Mainland China (1997)

MIR Name	Total area (Km²)			Nonagricultural population (million)		GDP (100 million Yuan)		Realized FDI (US$100 million)	
	Prefec-ture	Urban district	Built-up Area	Prefec-ture	Urban District	Prefec-ture	Urban district	Prefec-ture	Urban district
Middle-south of Liaoning MIR	77142	10038	885	13.4570	10.5552	2798.67	1987.13	20.82	19.03
Jing-jin-tang MIR	78265	12395	1147	16.8871	13.7509	4819.25	2758.27	44.32	10.86
Yangtze River Delta MIR	98591	10233	1289	27.5036	17.8069	11554.64	5499.13	115.35	83.91
Pearl River Delta MIR	54718	9768	694	10.8291	6.5307	5363.54	3442.17	91.56	55.48

Note: The data are for cities at and above the prefecture-level. The data for the Pearl River Delta do not include Hong Kong and Macao.

Data source: *Urban Statistical Yearbook of China, 1998*, Beijing: China Statistical Publishing House, 1999.

(1) Middle-south of Liaoning MIR: with Shenyang and Dalian as its cores, Shenyang, Fushun, Benxi, Liaoyang, Anshan, Yingkou, Panjin, Wafangdian and Dalian forming a massive MIR.
(2) Jing-jin-tang MIR: with Beijing and Tianjin as its cores, this massive MIR consists of two city rings — the inner ring (Beijing-Tianjin-Tangshan-Langfang) and the outer ring (Qinhuangdao-Chengde-Zhangjiakou-Baoding-Cangzhou).
(3) The Yangtze River Delta MIR: with Shanghai, Nanjing and Hang-zhou as its cores, Ning-shao Region (Ningbo and Shaoxing), Hang-jia-hu Region (Hangzhou, Jiaxing and Huzhou), Hu-su-xi-chang Region (Shanghai, Suzhou, Wuxi and Changzhou), Tong-tai Region (Nantong and Taizhou), Ning-zhen-yang Region (Nanjing, Zhenjiang and Yangzhou) and Ma-wu-tong Region (Maanshan, Wuhu and Tongling) joining together as a megalopolis.
(4) The Pearl River Delta MIR: with Guangzhou and Shenzhen as its cores, this massive MIR incorporates Shenzhen, Zhuhai, Dongguan, Foshan, Zhongshan, Jiangmen and Zhaoqing. Hong Kong and Macao can also be considered to be part of this MIR.

The above four MIRs are the most developed metropolitan areas in China whose central cities are either municipalities directly under the central government or at the vice-provincial level. Other metropolitan areas as listed below are also emerging:

(1) The region along the Jiao-ji and Jin-pu Railway and Jiaodong Peninsula: with Jinan and Qingdao as its cores, a zone with a high density of cities is being formed along the Jiao-ji and Jin-pu Railway, which includes Long-yan-wei-lai-qing Region (Longkou, Yantai, Weihai, Laiyang and Qingdao), Zi-qing-wei Region (Zibo, Qingzhou and Weifang), Ji-tai-wu-xin Region (Jinan, Taian, Laiwu and Xintai) and Ji-yan-qu Region (Jining, Yanzhou and Qufu).
(2) The southeast coastal areas of Fujian Province: with Fuzhou and Xiamen as its cores, along the coast a belt of high density of cities is emerging in the Fuzhou-Putian-Quanzhou-Xiamen-Zhangzhou region.

Furthermore, in the middle-west of China, some MAs are growing to varying degrees such as the Jianghan Plain with Wuhan as its core, Cheng-

Yu Region with Chengdu and Chongqing as its cores, Guanzhong Plain with Xian as its core, along the railway in the northwest of Henan Province with Zhengzhou as its core, the northeast of Hunan Province with Changsha, Zhuzhou and Xiangtan as its cores and the Songnen Plain with Harbin as its core, and so on.

Types of MAs

According to spatial organization and management patterns, we can classify MAs in China into three types.

(1) United Metropolitan Area (UMA): This kind of MA has a complete administrative division system with a super city as a central city, mainly municipalities directly under the central government, that is, Shanghai, Beijing, Tianjin and Chongqing. It has two characteristics. One is its great scope and the other is that it has a single center. The municipalities directly under the central government are the administrative units at the province-level, under which districts/counties but no municipalities are designated. In urban districts, the vertical administrative system is "municipality-district-street", namely "two-level governments, three-level management". In suburban districts/counties, the relevant administrative system is "municipality-county-town", namely "three-level governments, three-level management".

(2) Loose Metropolitan Area (LMA): In this kind of MA, with a super city or a big city as the center, there are many county-level municipalities (CLMs), where a "municipality governing municipality" (MGM) system is in operation. Most of these are the capital cities of a province and prefecture-level municipalities (PLMs) whose populations are smaller. Each has a non-agricultural population of more than one million. There are about 30 such MAs in the entire country. In such MAs, the county-level municipalities are mostly transformed from previous counties. CLMs are directly under the jurisdiction of a prefecture. Most have independent power of economic management, with a close vertical jurisdictional relationship with a PLM, but some only have a loose relationship with a PLM. Within these kinds of MAs, conflicts of interest are common.

(3) Independent Metropolitan Area (IMA): MAs of this type are comprised of two or more big cities at the same administrative level. Most of them are MIRs. They are concentrated in the areas with a booming economy and a high density of cities such as Shenyang-Dalian, Beijing-

Tianjin, Guangzhou-Shenzhen-Zhuhai, Changsha-Zhuzhou-Xiangtan regions and Ningbo-Shaoxing, Hangzhou-Jiaxing-Huzhou, Suzhou-Wuxi-Changzhou, Nanjing-Zhenjiang-Yangzhou regions in the Yangtze River Delta, and so on. Considering the administrative levels, each of such MAs has two to three independent prefecture-level municipalities with several municipalities (or counties) under their jurisdiction. Within such an MA, the prefecture-level municipalities are not far from each other, and they have similar population sizes. There are grievous conflicts of interest among prefecture-level municipalities or between a prefecture-level municipality and its county-level municipalities.

Compared with the MAs in foreign countries, MAs in China have their own outstanding characteristics. First, MAs include vast rural areas in addition to the urbanized areas. The rural population may constitute the majority of the population in the whole MA. Accordingly, the proportion of agricultural output in the value of GDP as a whole is fairly high. Second, MAs in China are mainly regions with a high density of cities where the "municipality governing municipality (or county)" system is in operation. There is a strong top-to-bottom administrative feature in these MAs. The administrative divisions result in rigid constraints on the development of a city or region in various aspects such as planning, construction and management. The city government not only is a local administrative body, but also strongly influences the economy and social development. Third, the present "jurisdictional economy" has greatly hindered the development of the market economy system, and obstructed economic integration in the MAs.

Problems of metropolitan administration

With the rapid development of the national economy and the rising level of urbanization, MAs in China are being formed and growing continuously. The current administrative divisions, organization and management system can no longer meet the needs of socio-economic development. Particularly at the present transitional stage from a planned economy to a market economy, enterprises have not completely established their independent operating status. The selfish departmentalism of local governments is very serious and local governments also have obvious economic behavior of their own. This has become a subjective obstacle for close cooperation and integrated development within MAs. Thus it is of utmost importance to reform the contemporary metropolitan administrative organization and management system. We should explore a proper way to meet the needs of economic and social development to solve various problems in MA management.

In view of the present situation, there are three types of problems in metropolitan administration in China.

Abnormal competition between cities with equal administrative status and economic strength

Cities of this type always have close historical, economic, social and cultural relationships and are joined with each other in space. Their levels of economic development are similar and usually host government organizations at the same level. For example, Suzhou, Wuxi and Changzhou in Jiangsu Province are all prefecture-level municipalities. Under the current urban administrative organization and management system, each city develops for the sake of its own local interests which is likely to result in local selfish departmentalism and utilitarianism. Thus, there are loose economic connections among the cities. Each develops within the scope of its own administrative jurisdiction and they do not open to each other. They have established a "small but complete" economy system. As a result, the distribution of industry is dispersed and industrial structures are duplicated. This constrains the expansion of the growth pole in regional economic development and the raising of the level of urbanization. The duplicated construction and abnormal competition among cities also leads to unnecessary wastes of limited resources. It obstructs the development of the regional economy to some extent (Liu and Wang, 2000; Liu, 1999; 2000).

Conflicts of interest between cities with a vertical administrative relationship

Though the "municipal governing county" (MGC) system carried out since the mid 1980s has enhanced economic linkages between urban and rural areas and the integrated development of the urban economy, more intense conflicts have emerged with the improvement in the economic strength of counties. There are various cases, such as Suzhou City and former Wu County (now Wuxian City), Wuxi City and former Wuxi County (now

Xishan City), Changzhou City and former Wujin County (now Wujin City), Guangzhou and Panyu, Foshan and Nanhai, and so on. On one hand, CLMs with the same seat as the PLM have selected and moved to towns on the urban fringe as their new centers in order to escape the administrative constraints from the PLM and to raise their administrative status. Thus, because of different interests between CLMs and PLMs, integrated plans cannot be carried out in the MAS, resulting in a division of urban services and infrastructure in the built-up areas. On the other hand, restricted by PLMs, those CLMs (or counties) with different seats from PLMs cannot expand the urban scope and infrastructure to meet the demands of economic development. Meanwhile, the economic growth of CLMs requires the promotion of their administrative status. However, for the sake of maintaining their own status, PLMs try to repress such attempts, which contributes to an aggravation of the conflicts between PLMs and CLMs and results in chaotic management.

Strange competition between cities without a vertical administrative relationship and cities of disparate economic strength

The built area of a cross-boundary metropolis covers more than two administrative units, for example, the cross-boundary MA involving the administrative areas of Haikou, Qiongshan and Chengmai. According to the general principles of economic development, the regional economic strategy should give priority to the construction of a central city in order to stimulate the development of the hinterland. It is inevitable that priority be given to the central city at the expense of the interest of the peripheral cities. For instance, the central city may not have a good harbor and needs to rely on the peripheral cities. It may also have limited space, restricting its further development, and it needs to annex or "borrow" land from the peripheral cities. It may face a shortage of water resources and needs a supply of water from other cities or conducts water supply projects in other cities. But all these harbors and lands are always located in the zone with the most favorable conditions and the most flourishing economy. In view of local interests, it is most likely that the peripheral cities would use these areas for their own use. This will restrict the development of the central city to a certain extent, resulting in conflicts between it and the peripheral cities. However, this will be inhibit the development strategy of the regional economy.

In fact, the contradictions in administrative organization and management in China's MAs are conflicts of various kinds of interest. Thus it is clear that in China the current economic, political and administrative systems, including the administrative divisions, have greatly affected the economic development of cities and regions. So it is necessary to reform the administrative system that does not fit the development situation in the MAs.

Innovation in the administrative organization and management model for China's MAs

Two types of metropolitan administration systems in the world

Up to now, China has not established its own system for administrative organization and management of MAs, except for the municipalities directly under the central government. Thus it is important to examine the experience of other countries' MAs in order to extract essentials from them for reference. With a history of several decades of development under particular political and cultural circumstances, there are two models of metropolitan administration in western countries.

(1) Metropolitan public organization under a single-center system: The single-center system, also called a unitary system, has only one policy-making center and an integrated management body in an MA. It is obvious that this system provides the proper organizational size for public services, such as a harbor, airport and other communication facilities and water diversion works. This type of system can reduce competition and conflicts which may obstruct the development of MAs. It can also expedite the flow of resources and effectively link together community services belonging to different administrative bodies. Thus, the MAs will achieve an economy of scale. However, there are other views and doubts about this system. For instance, if the size of the public institution is too large, it may not represent the interests of all sides and meet all kinds of needs. The costs for controlling may be high which may reduce the effectiveness of the institutions. Particularly, the single-center system is apt to fall into a structural crisis owing to its hierarchical bureaucracy. The evident manifestations are that the government is slow in reacting to the citizens' daily needs and it does not react on behalf of local public interest. Furthermore, the single-center system leads to a rise in administrative costs and a reduction in welfare due

to the shortage of competition. According to the analysis above, we know that public organizations under a single-center system can only provide limited community services in MAs.

(2) Metropolitan public organization under a polycentric system: The polycentric system, also called polyatomic system, has more than one policy-making center, including the formal and comprehensive government units (state, city, town, and so on) and many particular areas (school districts or non-school districts) overlapping with one another. In western countries, particularly in America, this kind of system is very common. Particular divisional organizations other than school districts grow rapidly. All the divisions and changes in the administrative districts and the foundation of the coordinate organizations are the results of the pursuit of the economic interests of specific community services. Thus a polycentric system tries to meet the various demands and appetites of the citizens. Due to the small scale of the government, the public can easily participate in the monitoring. Therefore, the government is sensitive to changes in the citizens' demands. The main problem with a polycentric system is how to realize the public interest in a wider scope beyond various kinds of minor functional districts. Only through the cooperation, competition and consultation of various local units can large-scale community services be realized. Cooperation will be easy to realize if every related side can represent the public interest completely. The united act will bring great benefits to all sides. In fact, cooperation is very difficult to realize because in many cases the distribution of the consumption and benefits of the public institutions and services is not even. There are conflicts about the costs and interests among the related sides. But owing to the polycentric system, such conflicts can be resolved, and at the same time reasonable competition among the sides will be reserved. But under the single-center system, every local government has its own veto power, so it is not easy to take a united action in an MA.

A "metropolitan coalition," or the governmental united organization, as a metropolitan organizational model created by the western countries provides references for our country. With the different political, economic, human and physical circumstances and the following differences in the citizens' appetites, various countries have chosen different models for public organization and management in MAs. Some have established a metropolitan government, and some have established jurisdictional districts with different functions and overlapping scales. It should be emphasized that these jurisdictional districts with specific functions play important roles in metropolitan public organizations. The function of such districts is always unitary, such as education, environment protection, flood control, fire service, providing public facilities including public traffic, street lighting, health care, funeral and internment services, and so on. Such districts are needed for metropolitan development and to meet the various demands of the residents. In many cases, these particular districts can be managed properly, obeying the rules of the scale economy in order to achieve better economic and social benefits and to alleviate the burdens of the government.

In light of related laws, reforms of the public administrative organization and management in the West consistently give top priority to the public interest. Thus public participation is emphasized. As the general representation of local interests, through proper intervention, the urban government can overcome economic and social problems resulting from a complete market economy. However, intervention is appropriate and limited. The government administers the economy by way of laws, taxation and private property management rather than by direct intervention. Modifications of the government jurisdiction are rigorously carried out according to legal procedures. In view of the public interest, it is unclear whether the single-center system or the polycentric system is better. Each has its own background, advantages and disadvantages. It is generally agreed that under a polycentric system there are many organizations of different scales that can provide the best organizations for public goods production and consumption and all kinds of community services. In contrast to the single-center system, it can meet public demands better. It also enhances the relevance and practicality of the government's public administration.

Innovation in administrative organization and management in Chinese MAs

(1) Fundamental principles for innovation: Following the principles of management science, economics and urban development, and taking into consideration the situation of our country and the result of a comprehensive analysis of the present situation and problems, and drawing on the experiences and lessons from the West, we propose the following principles for an administrative reform in our MAs:

  a. Integration of efficient administration and rapid economic development, i.e., the new metropolitan administration system must provide circumstances for the fastest economic development and the most effective administration.

b. Unity between general interests and local interests, that is to say, within an MA, we should consider both the general interests of the MA and the local interests of individual cities within the MA.

c. Linking scientific considerations and feasibility together, which means that the reform scheme should not only be scientific and reasonable, but also operable with results.

d. Integration of pursuing the benefits of the scale economy and sharing of benefits. In view of the aim to provide more community services for residents, it is important for the metropolitan administrative organizations to seek a scale economy. At the same time, the allocation of benefits must be solved properly. Only in this way can the reform programs and measures be workable and will there be as few new contradictions as possible.

(2) Proposal for the innovative model: Different from the western state system and form of government, there are great differences in the quality and content of metropolitan administration and the emerging problems between our country and the western countries. In view of the situation in our country, drawing lessons from the West and having the settlement of the administrative problems in mind, I propose the following suggestions:

a. To establish a highly centralized Metropolitan Government (MG), namely, to build a kind of administrative organization at a level between the province and the city by means of coalition or annexation. The advantages of this are as follows. First, the framework of a highly centralized MG is beneficial for prompt implementation of policies. Second, it is favorable for integrated metropolitan planning. Thus the resources and capital of various cities can be fully used. Integrating community service projects of all the cities, the services can achieve a scale economy to meet the needs of residents outside the boundary. Meanwhile, there are also disadvantages. The establishment of a MG means the addition of a number of administrative organs, which will lead to a reduction in administrative efficiency and new administrative interventions to the economy. And it runs counter to the current policy that is to simplify the administrative structure. The centralized MG may tend to stress local interests and it may not represent all sides' interests. Thus the development of some cities would be restricted and the various demands and appetites of the residents would be neglected. New conflicts within the MAs may also emerge. Moreover, the over-centralized urban functions may neglect the grassroots, contributing to blindness in policy-making and a low efficiency of administration.

b. To establish a loosely integrated coordinating organization (non-government organization). As it is not easy for MAs to carry out their cross-boundary functions comprehensively, some non-governmental integrated organizations should be responsible for cross-boundary functions. There are several advantages. First, it will result in efficiency for some public services and such organizations can meet the demands of residents effectively. Second, such small-scale organizations under the residents' monitoring are sensitive to their demands, which strengthens the transparency and the relevance of decision-making. Third, the nature of non-government guarantees flexibility and sound metabolism, but it weakens the coordinating functions of cross-boundary public services in a greater region. Under the present circumstances of our country, the result of the imple-mentation of any decision is uncertain without direct administrative intervention. Thus the benefits of the scale economy of community services will be greatly reduced. Furthermore, if not properly handled to produce effective results, most probably the integrated organizations will have to be disbanded. Thus MAs will return to the same old disastrous road, even deepening the previous predicament. A typical case is the Shanghai Economic Region, established with the support of the State Council, which was forced to disband several years later.

c. To establish a united government with only cross-boundary functions. To resolve the current situation that MAs cannot perform common cross-boundary functions, an alternative is to establish a united government with only cross-boundary functions. This can harmonize the interests among local governments and solve the problems of public services among them. We can also call it a "Metropolitan Union (MU)". It incorporates the advantages of the above two schemes. It would meet the over-boundary needs of the residents and their various demands without restricting the local governments' exercise of their non-cross-boundary power. Moreover, it not only retains some administrative interventions, but also prevents the formation of overstaffed administrative organs. Although a MU is different from a centralized MG and it will have fewer negative effects, some of a MU's functional organs still have some disadvantages, similar to those of a MG to a certain extent.

The three models above each has their advantages and disadvantages. We should consider local conditions to choose a proper model. Generally, the third model is preferable and is the most suitable for the current reform of metropolitan administration in China.

Administrative Division Reform in the Pearl River Delta

Characteristics of the Pearl River Delta MIR

The Pearl River Delta MIR is a typical MIR in our country. It includes 2 vice-province-level municipalities, 7 prefecture-level municipalities, 16 county-level municipalities, 3 counties and 22 municipal districts (see Table 2). It has an area of 41,600 square kilometers, a *hukou* population of 21 million, an urban population of 8.8 million, and a temporary population of 6 million. With a GDP of more than 230 billion yuan in 1994 and occupying 22% of the area of the whole province, it is one of the most developed economies in the region, with a high population density and a concentration of various cities and towns. Its level of urbanization reaches about 45%, 14 and 22 percentage points above the average of Guangdong Province and the whole country respectively (Group for Metropolitan Planning of Economic Region of the Pearl River Delta, 1996: 11). Since 1979, benefiting from the open policy and its advantageous location near Hong Kong and Macao, the Pearl River Delta has achieved rapid economic development. From 1979 to 1994, the average rate of GDP increased over 20% annually. Its industrial structure was upgraded in an export-oriented economy. As a result, the levels of infrastructure, basic industries and the living standard of the people have improved enormously. The scope of the built-up areas has expanded rapidly.

There are notable characteristics in the economic development of the Pearl River Delta. (a) Economic development has been very rapid. (b) A large share of the economy is export-oriented. (c) The gap between the urban and rural areas has been reduced to a great extent. (d) The unique boundary location close to Hong Kong and Macao is an important

Table 2. Administrative divisions in the Pearl River Delta (1998)

Municipality	Municipal district, County-level municipality or County	Total
Guangzhou	Dongshan district, Liwan district, Yuexiu district, Haizhu district, Tianhe district, Fangcun district, Baiyun district, Huangpu district, Panyu municipality, Conghua municipality, Huadu municipality, Zengcheng municipality	8 districts and 4 county-level municipalities
Shenzhen	Luohu district, Futian district, Nanshan district, Baoan district, Longgang district, Yantian district	6 districts
Zhuhai	Xiangzhou district, Doumen county	1 districts and 1 county
Foshan	Chengqu district, Shiwan district, Shunde municipality, Nanhai municipality, Sanshui municipality, Gaoming municipality	2 districts and 4 municipalities
Jiangmen	Pengjiang district, Jianghai district, Taishan municipality, Xinhui municipality, Kaiping municipality, Heshan municipality, Enping municipality	2 districts and 5 municipalities
Huizhou	Huicheng district, Huiyang municipality, Boluo county, Huidong	1 district, 2 countries and 1 municipality
Zhaoqing	Duanzhou district, Dinghu district, Gaoyao municipality, Sihui municipality	2 districts and 2 municipalities
Dongguan		
Zhongshan		

Editors' note: Guangzhou annexed Panyu and Huadu as two districts in 2000. Zhuhai annexed Doumen as a district in 2001.

contributing external factor to the rapid economic growth of the region, particularly the role of the Special Administrative Region of Hong Kong which implements the "one country, two systems". (e) Since the 1990s, tertiary industry has increased, keeping pace with manufacturing development which began to diffuse to the rural areas. There is a clear trend of integration of the urban and rural areas.

The Pearl River Delta has a high urbanization level, a relatively powerful economy and also advanced transportation and communications facilities. A complete urban network system has been formed in this region.

In addition to Hong Kong and Macao, there is a super city (Guangzhou), a big city (Shenzhen), 12 medium cities (with populations of 250,000 to 500,000), 11 small cities (with populations of less than 250,000) and several hundred designated towns. Among them, Guangzhou and Hong Kong are the super economic centres, followed by Shenzhen at the second level and Zhuhai, Macao and other medium and small cities at the third level. Meanwhile, Guangzhou, Shenzhen and Zhuhai are the most important economic growth centers in the MIR. There are three MAs, with Guangzhou, (Hong Kong) Shenzhen, (Macao) and Zhuhai as the respective cores which lie respectively in the middle, east and west. The MIR takes the Guang-Shen-Gang (Guangzhou, Shenzhen and Hong Kong) and Guang-Zhu-Ao (Guangzhou, Zhuhai and Macao) zones as its developmental axes.

The metropolitan administration system of the Pearl River Delta is very complicated with many changes. On the mainland side of this region, there are 2 vice-province-level municipalities, 7 prefecture-level munici-palities, 16 county-level municipalities and 3 counties. There are five types of prefecture-level municipalities as follows. (a) Vice-province-level municipalities with strong economic power, namely Guangzhou and Shenzhen. In Guangzhou, the "municipality governing districts" and "municipality governing municipality" systems have been implemented. There are 8 districts and 4 municipalities (Panyu, Conghua, Huadu and Zengcheng) under its jurisdiction. There are 6 districts under the jurisdiction of Shenzhen. (b) Prefecture-level municipalities with a specific economic zone with relatively strong economic power, i.e., Zhuhai. Under its jurisdiction there is one district and one county. (c) Prefecture-level municipalities with small gaps in economic strength between the central cities and the peripheral cities, such as Foshan and Jiangmen which administer 2 districts, 4 municipalities and 2 districts, 5 municipalities respectively. (d) Prefecture-level municipalities whose economic strength is relatively weak. They are the results of a coalition of the prefecture administrative office and the county-level municipality. Huizhou and Zhaoqing are in this category with one district, one municipality and two districts, two municipalities within the Pearl River Delta respectively. (e) Prefecture-level municipalities upgraded from county-level municipalities with no municipalities, counties and districts under their jurisdictions, such as Zhongshan and Dongguan.

Problems of the current metropolitan administrative system

The current pattern of administrative divisions in the Pearl River Delta is not only the result of history, but also the result of urban or regional development. So it has its own necessity and rationality to a certain extent. It contributed to the economic development of the Pearl River Delta. Basically it can meet the needs of economic development during the specific period. Nevertheless, in the new period into the 21st century, the present administrative system will become an inhibiting factor to development as it does not fit the sustainable development and integrated planning and construction of the MIR (Wei and Yangluo, 1997; Ye *et al.*, 1999). The main problems are as following.

(1) The administrative levels are excessive and chaotic with no formal management: As mentioned above, there are five types of administrative division systems among nine prefecture-level municipalities. This is not very normal. Particularly, the operation of the "municipality governing municipality" system has added a new administrative layer and expenses. The distance between the leaders and the people has been increased, which has encouraged the bureaucracy and lowered the efficiency of the administration. Furthermore it is very harmful to the effective implementation of socialist political democratization.

(2) The wide area urbanization model has confused the concepts of "city proper" (built-up area) and "urban area" (area under a municipality's jurisdiction): This has fostered the bubble of land development, resulting in out of control use of land for construction. The current model encourages the blind development of land to some extent. Some data show that from 1990 to 1993, due to excessive land sales without planning, the land used for urban construction in the Pearl River Delta has tripled (Group for Metropolitan Planning of Economic Region of the Pearl River Delta, 1996: 15). All local areas are eager to widen the planned population and the size of the urban land. Thus the waste of the land becomes serious and the environment pollution is aggravated.

(3) The "designation of county as municipality" policy has strengthened the power of the county-level municipality and mobilized its initiative: But the macro control ability has been reduced. The development of the central cities in the Pearl River Delta is confined by space and some policies. These cities cannot realize systematic divisions of labour with more competition than cooperation among them. Each wants to be big and complete. This

has hindered the harmonious development of the cities and the region. Thus, it is very difficult to draw up an integrated plan and implement united construction and administration.

(4) Each city does its own infrastructure construction: In order to improve its status in the MIR, many cities, especially the county-level municipalities with relatively strong economic strength, pursued the construction of large and complete infrastructure and community service facilities. Each has its own system. So there is much duplicated construction, such as airports, harbors and roads. There is a distinct disparity between the transportation capacity and the actual transportation flow. This results in a massive waste of financial resources. Meanwhile, the construction of transport and communication facilities between cities has been neglected. This has negatively influenced the strength of the Pearl River Delta.

It can be concluded that the existing system no longer fits the sustainable development, and reform is urgently required.

Considerations of administrative division reform in the future

(1) Principles for administrative division reform: The principles for the reform of the Pearl River Delta are: priority for the interest of the region as a whole, facilitating development, easy management with few administrative layers, experiment first with smooth reform, adaptable to the local conditions, consideration of public opinion. The aim of the reform is to provide a good institutional environment to propel the development of the metropolitan area in the Pearl River Delta.

There are three main bases for the reform. One is the relevant laws and regulations on administrative divisions. The second is the plan of the economic region in the Pearl River Delta including the plan for a city system, infrastructure, industry location, environmental protection and social development. The third is the main conflicts about the existing administrative system.

(2) Reform options: (a) To establish a metropolitan government union with cross-boundary responsibilities. A metropolitan union is a semi-administrative government, whose responsibilities only concern the cross-boundary functions. For example, to organize and implement important cross-boundary projects in the fields of infrastructure, such as drainage, water supply, transport, communication, energy and environmental protection. The union consists of nine municipalities, with one official at

the sub-province level as chairman, and the mayors from Guangzhou, Shenzhen and Zhuhai as vice-chairmen.

(b) To set up three coordinating committees under the metropolitan union in accordance with the plan of the future urban system in the Pearl River Delta. Guangzhou metropolitan area is in the middle part, centered in Guangzhou, including Foshan, Nanhai, Sanshui, Huadu, Panyu, Conghua, Zengcheng, Shunde, Gaoyao and three districts of Zhaoqing. The east is the Shengang MA with (Hong Kong) and Shenzhen as its centers, including Dongguan, Huizhou, Huiyang, Huidong and Boluo. The west is the Zhuao MA, centered on (Macao) and Zhuhai, including Doumen, Zhongshan, Jiangmen, Xinhui, Heshan, Gaoming, Taishan, Kaiping and Enping. The mayors from Guangzhou, Shenzhen and Zhuhai will be the directors of each committee. Their duty is to make the plan for the MA, to coordinate the interests, and to carry out all tasks from the metropolitan union.

(c) To establish a municipal hierarchy instead of the "municipality governing municipality" system. In view of the many conflicts in the Pearl River Delta, the existing system should be abolished to re-group the municipalities in MAs with relevant administrative organization. A municipality hierarchy system should be implemented according to local conditions, population and economy. The administrative hierarchy can be kept for a certain time, but all municipalities should be equal in political power.

(d) To establish more municipalities under the central government (MUC). The MUC plays an important role in the fields of the politics and economy of mainland China. Now the few MUCs in China are not distributed evenly. So it is necessary to establish more MUCs. South China is one of the economic centers with two super-metropolises (Guangzhou and Shenzhen) at the sub-province level. Either is competent to be upgraded to a MUC. In view of population, economy and geographical location, Guangzhou has more advantages over Shenzhen. But in view of the degree of openness, industrial globalization and urban management, Shenzhen is more competent than Guangzhou. However, as the capital city of Guangdong, it is more difficult for Guangzhou to become a MUC, which contradicts the integrated plan, construction and management in the Pearl River Delta. In contrast, it is easier for Shenzhen to be upgraded to a MUC. It is especially helpful to strengthen the cooperation with Hong Kong and then to form the Gang-Shen MA. If it is upgraded to a MUC, Shenzhen should be subordinated to the integrated plan of the whole metropolitan

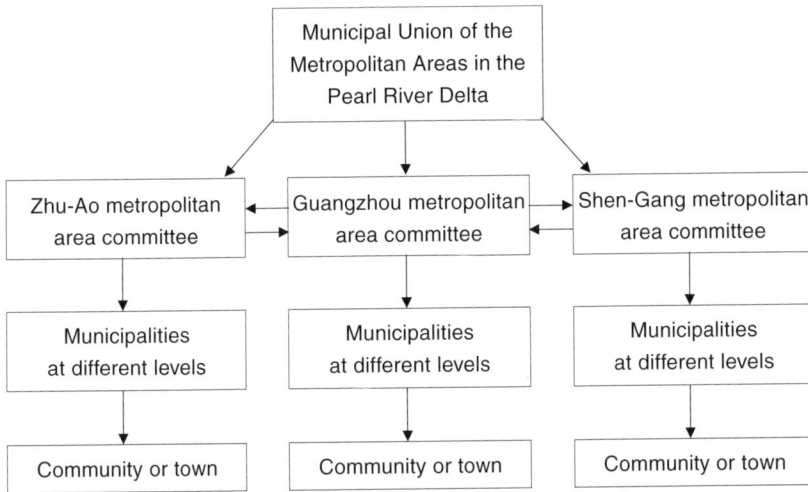

Figure 1. A model of a vertical administration system after reform

area in the Pearl River Delta. The model of a vertical administration system after reform is shown in Figure 1.

Conclusion

Some conclusions can be obtained. First, after reform and opening, there has been a great development of Chinese MAs. The problems of administrative organization and management in the MAs have attracted attention from researchers and governments. Thus, the study of the administrative system in MAs in accordance with Chinese conditions is important both in practice and theory.

Second, in view of the administrative divisions, there are three categories of MAs: united MAs, loose MAs and independent MAs. The latter types implement the "municipality governing municipality" system which has many conflicts and serious problems of integration. Currently the organization model has seriously constrained the integrated interests of MAs and the region. It is also the key to Chinese administrative reform.

Third, the administrative model in foreign countries has developed for several decades. There are two models, *i.e.*, a single-center model and a polycentric model. Public services are organized in different ways. Different governments make different choices about public administrative organization. Some establish metropolitan governments, while others establish various administrative districts with different functions or sizes that can be overlapping. The experiences of foreign metropolitan organizations and administration can be used for reference for the reform in China.

Fourth, there are three options for metropolitan organization and administration in China. One is to establish a centralized metropolitan government. The second is to establish a loose coordinating organization for MAs and the third is to establish a cross-boundary united government. Each model has its advantages and disadvantages. Considering Chinese conditions, the third option is more suitable than the other two.

Fifth, the Pearl River Delta is one of four MIRs in China. The existing administrative system is no longer suitable to the integrated development of the MIR. Reform is inevitable. A possible reform is tò establish a government union for the MAs in the nature of a semi-government, to set up three committees of MAs under a government union (Guangzhou/Shen-Gong/Zhu-Ao), to implement a municipality hierarchy instead of a MGM, and to establish more MUCs.

Sixth, administrative divisions and the management system are complicated topics. They demand cooperation between researchers and government. In terms of the conflicts in the Pearl River Delta, it is urgent that a new system be instituted. Moreover, it is also feasible to conduct experimental reforms in the Pearl River Delta.

References

Group for Metropolitan Planning of Economic Region of the Pearl River Delta. *The Metropolitan Plan for the Economic Region of the Pearl River Delta.* Beijing: Chinese Construction Press, 1996.

Liu Junde. "A long-term neglected field — cross-boundary administration and organization." *Journal of Hangzhou Normal College*, 1 (1999).

Liu Junde. "Study of integration of the spatial economy in the Yangtze Delta." *Journal of Hangzhou Normal College*, 1 (2000).

Liu Junde and Wang Yuming. *System and Innovation—New Issues about the Development and Reform of the Chinese Municipality System.* Nanjing: Dongnan University Press, 2000.

Wei Qingquan and Yangluo Guancui. *Administrative Division of the Pearl River Delta at the Turn of the New Century.* Guangzhou: Guangdong Map Publishing House, 1997.

Ye S., Niu Y. and Gu C. (eds.). *Studies on Regional Integration under the Model of "One Country, Two Systems"*. Beijing: Sciences Press, 1999.

PART V

Concluding Remarks

Studies of the Pearl River Delta: New Findings and Future Research Agenda

Alvin Y. So

School of Humanities and Social Science,
Hong Kong University of Science and Technology

This volume is a collection of very interesting papers on the Pearl River Delta (PRD) region. In this concluding chapter, I will first summarize some of the key findings in this volume, highlight their contributions to the PRD literature, and spell out their policy implications for the future development of the region. Then I will raise a few methodological issues and try to outline the future agenda for PRD studies.

Since I am a sociologist by training and I know very little about resource management, my following discussion tends to focus on issues relating to spatial differentiation in the PRD, regional integration, urban migration and urban governance.

Causes of spatial differentiation in the Delta

What are the factors that explain the rapid economic growth and urbanization in the PRD over the past two decades? Has rapid economic growth led to significant changes in the PRD's spatial structure? Using a decomposition approach based on regional production functions, Shen *et al.*'s analysis of the data provided by the Guangdong Statistical Bureau attempts to answer the above two questions.

The significance of local investment

Shen *et al.* find that local investment has been the most important contributor to economic growth. During the 1990–1994 period, 45 percent of the economic growth in the PRD was explained by local investment. This figure is much higher than the 27 percent figure for technical progress, the 17 percent figure for non-local investment (mostly from Hong Kong), the 9 percent figure for an increase in the labour force, and the 2 percent figure for the scale economy. Thus, although the PRD is proud to have attracted a very large amount of non-local investment from Hong Kong, Shen *et al.* show that it is local investment that has played the most significant role in explaining the economic growth of the region.

Shen *et al.* further point out that even in the two special economic zones of Shenzhen and Zhuhai, non-local investment from Hong Kong contributed only about 18 percent of economic growth during the 1980–1994 period. In fact, during the initial 1980–1985 period, local investment accounted for almost 70 percent of economic growth in these two cities.

Since non-local investment from Hong Kong is often perceived to be the crucial factor in the so-called PRD miracle, Shen *et al.*'s findings provide us with some quantitative evidence to challenge this commonly-accepted explanation. If local investment was indeed the most crucial factor, researchers will need to reconsider domestic factors so as to account for the rapid development of the PRD. Future research projects may show that in the absence of the mobilization of capital and labour by local state agencies and Guangdong entrepreneurs, the PRD would have had difficulties in achieving such a rapid rate of development over the past two decades.

The rise of Shenzhen

Another significant finding of Shen *et al.* is the rise of Shenzhen and the fall of Guangzhou in the PRD's urban hierarchy. The GDP share of Guangzhou declined rapidly from 43 percent in 1980 to 24 percent in 1990, and it declined further to 22 percent in 1998. On the other hand, the GDP share of Shenzhen in the region increased dramatically from only 2 percent in 1980 to 14 percent in 1990, and it increased further to 22 percent in 1998. By 1998, Shenzhen's GDP share was very close to that of Guangzhou and it was significantly greater than that of other localities in the region.

By the late 1990s, a new spatial economic structure, with two economic centers, had emerged in the PRD. In 1998, the total GDP share of Guangzhou and Shenzhen in the region was as high as 44 percent, while Dongguan (the third largest economic center) accounted for only 6 percent.

Chu *et al.* also point out that Shenzhen is growing at a rate beyond the imagination of Hong Kong observers. The double-digit growth rate over the past three decades has transformed Shenzhen from a small frontier town to the sixth largest urban economy in mainland China. The population of Shenzhen is highly educated, and it has been estimated that 20 percent of the Ph.D.s in China are found in Shenzhen. The gaps in housing prices and wage levels between Hong Kong and Shenzhen are narrowing due to the rising productivity in Shenzhen. It is projected that by 2005 Shenzhen's per capita GDP will be US$15,000, and that Shenzhen will catch up with Hong Kong in per capita output in less than 20 years. Thus, it is incorrect to regard Shenzhen as a satellite town of Hong Kong.

The rise of Shenzhen will have a profound implication on Hong Kong's integration with the PRD.

Regional integration

The Basic Law spells out the relationship between the Hong Kong Special Administrative Region (SAR) and the central government in Beijing. However, the Basic Law says very little about the relationship between Hong Kong and other Chinese cities or about the role of Hong Kong in the PRD region. As such, the Hong Kong SAR government is free to pursue its own pattern of regional integration with the PRD. For some reason, a "super-fortress Hong Kong" mentality has prevailed in the SAR, with public opinion in favor of stricter controls on mainland migration to Hong Kong.

Against the policy of a "Super-fortress Hong Kong"

Despite the enormous amount of investments from Hong Kong in the PRD, Sung shows that there are many barriers to the economic integration of the two regions. For instance, it is impossible for Hong Kong and the PRD to form a formal trade bloc on the model of the European Union. Even the establishment of a border industrial park is fraught with difficulties.

Still, Sung is against the policy of a super-fortress Hong Kong. Although Hong Kong public opinion favors stricter controls, not only for family reunions but also for children who have constitutional rights of abode in Hong Kong, Sung argues that Hong Kong's stringent controls on migration have to be relaxed. The ultimate solution to population pressures in Hong Kong lies in deeper integration with the PRD. Hong Kong can encourage people to relocate in Shenzhen by expanding the capacity of border crossings to facilitate commuting. Hong Kong can also encourage retired people to relocate in the PRD by building housing estates there designed for retirees.

The reversion of Hong Kong to China should have facilitated policy coordination between Hong Kong and mainland China. But Sung criticizes that the present coordination is quite ad hoc and reactive, lacking in overall vision and strategy. Since Hong Kong's long-term interest is in closer integration with mainland China, Sung advocates that it is important to have better policy coordination on border area issues, regional infrastructure, tourism, technology policy, financial markets, stabilization of the exchange rate, and China's entry into the WTO.

Toward state initiatives

The chapter by Gu *et al.* raises the issue of regional integration to the higher level of industrial upgrading and globalization. Since the 1980s, Hong Kong has been undergoing a process of industrial upgrading from labour-intensive industries to informative-intensive and high-tech industries. In merely a decade, Hong Kong has transformed itself from an industrial city to a global service center, providing such vital services as finance, trading, insurance, entrepot and transshipment, and R&D to the PRD, mainland China, and the East Asian region.

Gu *et al.* point out that against the backdrop of globalization and industrial restructuring, Shenzhen's municipal government also plans to upgrade its industry from labour-intensive and low-value added production to capital-intensive and high-tech production. The recent formation of an "Eastern Industrial Group", which is aimed at attracting high-tech investment to the Longgang region, aptly illustrates the determination of the Shenzhen government toward this goal.

Globalization and high-tech production require different types of actors to promote economic integration between Hong Kong and the PRD. During the previous period of labour-intensive production, small and medium (S&M) firms were the main actors and "sanlai yibu" (commissioned processing and compensation trade) was the dominant form of regional integration. However, in the present period of high-tech production, the S&M can no longer play a dominant role in economic integration. Instead, the governments in Hong Kong, Shenzhen, and other PRD localities have to take a much more active role in bringing large transnational firms together in various high-tech ventures.

Partnership between Hong Kong and Shenzhen

The high-tech era also has transformed the relationship between Hong Kong and Shenzhen. In the 1980s and the 1990s, Shenzhen was regarded as a satellite town of Hong Kong. In the early 21st century, however, Chu *et al.* argue that the gap between Hong Kong and Shenzhen has been narrowing. With a profit tax rate of 15 percent (Hong Kong's is 16.5 percent), with many incentives for high-tech industries, with strong government encouragement and a plentiful supply of skilled engineers and technicians, Shenzhen is very attractive to Hong Kong entrepreneurs. After the introduction of young, educated, open-minded cadres into key civil servant positions, the Shenzhen government has widely adopted international practices of doing business. In some aspects, like urban greening and tree planting along important roads, Shenzhen is even superior to Hong Kong.

Chu *et al.* contend that as a result Shenzhen should be seen as a partner, not a satellite, of Hong Kong. The relationship between Hong Kong and Shenzhen is symbiotic. Taking a longer perspective, Hong Kong-Shenzhen will become a twin-city in ten years' time, and the prosperity of Hong Kong will be dependent on Shenzhen's rapid economic growth.

Aside from questions of differential spatial development and regional integration, the chapters in this volume also examine issues of urban migration and urban governance.

Urban migration

Urban enclaves and social polarization

Economic restructuring in China has been accompanied by the mass migration of peasants into the cities and the emergence of migrant settlements. In Guangzhou, the number of temporary residents is estimated to be 1.3 million, 80 percent of whom live on the urban fringe. These temporary migrants live in *chu zu wu*, i.e., apartments and houses for lease, mainly built by farmers on their remaining collective land in the villages.

Taubmann reports that the living conditions of temporary migrants are abominable. Located on the urban fringe, many villages that rent out c*hu zu wu* still lack urban infrastructure facilities such as road networks, electricity, water supplies, drainage and sewage systems, fire prevention lanes, basic urban services, and especially sanitary services. The hygienic situation is poor, streets do not have asphalt surfaces, and fire risks are high.

In addition, Taubmann points to the social polarization between temporary migrants and local residents. On the one hand, temporary migrants are regarded as outsiders and excluded from the local society. On the other hand, temporary migrants have developed subcultures of their own. They communicate primarily with people from their native regions, speak their native dialects, maintain their own eating habits, live in closed communities, and share strong feelings with their home districts.

Tensions are growing between the temporary migrants and local residents. The local residents assign many negative labels to the migrant enclaves, such as "a paradise of thieves and robbers", "camps for prostitutes", "retreats of hunted criminals", and "typhoon harbours for floating people escaping governmental control over family planning". Social order has become the "number one public enemy" of local residents, and temporary migrants are the "root cause" of local residents' feeling of insecurity.

Segmented labour markets

Fan's chapter approaches the topic of migration from the perspective of segmented labour markets. She distinguishes two types of migration which match with two types of labour market. The first type is "state-sponsored migration" which consists of *hukou* (household registration) migrants from an urban background. *Hukou* migrants are highly educated. Their migration is sponsored by the state because they migrate either for the reason of job transfer or educational attainment. Not only do *hukou* migrants have the right to stay in the cities, but they also have access to high-paying jobs in the primary labour market which provides health care and retirement benefits.

The second type is "self-initiated migration" which consists of non-*hukou* migrants from rural backgrounds. Non-*hukou* migrants are poorly educated. Their migration is self-initiated because it is not supported by the state. They have no rights to stay in the city, and their migration is supposedly temporary, unofficial, and "outside of the state plan". Dislocated from institutional resources, they can only enter the secondary labour market and take jobs that pay low salaries and provide few benefits other than lodging.

Fan has convincingly shown that Chinese urban labour markets are segmented between *hukou* and *non-hukou* migrants. This finding, together with Taubmann's previous finding, will have significant implications for the study of political stability in the PRD. Temporal non-*hukou* migrants, because of their inferior legal status and poor human capital, clearly belong to a disadvantaged group. They are denied a chance to enter the primary labour market; they have to work long hours but they receive low wages and few benefits; they live in the less desirable *chu zu wu*; they form a migrant community; they are labeled as outsiders and identified as trouble makers; and tensions are growing between them and local residents. It is likely that future political conflict will erupt along the fault line between temporal migrants and local residents. In this respect, migration issues should be the focus of attention of any study of urban governance in the PRD.

Urban governance

Toward a corporatist mode

Xu's chapter argues that urban governance in mainland China is different from that in Western countries. This is because of different historical backgrounds, cultural traditions, and environments.

Xu argues that Chinese urban governance has the following four characteristics: state-led, multiple participation, democratic consultation, and scientific decision-making. Chinese urban governance has to be state-led because the state plays an indispensable role in macro economic control, providing public services and commodities, and supervising market operations. Multiple participation is necessary because the Chinese government always pays attention to public opinion and follows the mass line. Democratic consultation in China takes many forms, including public consultation, listening to expert opinions, holding seminar discussions, and interviewing concerned citizens. After undergoing the above processes, the decisions reached will be scientific because they are made after multiple participation and democratic consultation have taken place, and because they take into account the long-term economic, social, and environmental benefits of the society.

If we use the terminology of Western political scientists to "deconstruct" Xu's argument, we can say that Xu is trying to articulate a "corporatist" mode of urban governance for the Chinese government. Different from the Western democratic mode of urban governance which stresses elections, party competition, and the rights of public protests outside the state, the Chinese "corporatist" mode is aimed at absorbing societal conflict into the state and resolving it within the state. In the "corporatist" mode, the state, not the civil society, is the dominant institution, and the state defines the legal channels and the scope of appropriate behaviour by which societal forces are allowed to articulate their interests. Of course, societal forces can criticize the state and challenge the state's decisions, but they are permitted to do so only through the existing legal channels. Since societal conflict is completely absorbed within the existing legal channels, it would appear that there is an absence of open street protests, public anti-state activities, and violent confrontations between the state and societal forces.

Critics from the Western democratic tradition may dismiss Xu's "corporate" mode of urban governance as hypocritical because there are no true elections or party competition in China. But Xu's concrete examples show that Chinese local governments do take the opinions of migrants, local residents, and interest groups seriously, and that the governments do modify their policies in order to meet part (if not all) of the demands of the societal forces. If the Chinese government does have a corporate mode of urban governance, this may help to explain why China is able to maintain political stability even while its economy is undergoing such profound structural transformations that lay the groundwork for intensive conflict in the society.

Problems of Metropolitan Administration

Another characteristic of urban governance in China is the overlapping of metropolitan administration, which has resulted in the duplication of infrastructure facilities.

Liu reports that in the PRD there are two vice provincial-level municipalities, seven prefecture-level municipalities, 16 county-level municipalities, and three counties. With such a complicated structure of metropolitan administration, urban governance in the PRD is chaotic. Each municipal government wants to be big and to upgrade its status in the urban hierarchy. Thus county-level municipalities want to turn themselves into prefecture-level municipalities, and prefecture-level municipalities want to become provincial-level municipalities. They all want to have their own infrastructure and community service facilities. The end product of this chaotic urban governance is five international airports (Hong Kong, Macau, Shenzhen, Zhuhai and Guangzhou), four local airports, and many local harbours, leading to excess capacity, duplication of facilities, massive wastes of resources, and intensive competition among local governments.

To solve this problem, Liu suggests regrouping the municipalities into three areas. (1) The middle part is Guangzhou Metropolitan Area centered in Guangzhou. (2) The eastern part is Shengang Area centered around Hong Kong and Shenzhen. (3) The western part is Zhou'ao Area centered around Macau and Zhuhai.

Methodological issues and future research agenda

The chapters in this volume not only provide many significant findings for urban planners and policy makers, but they also raise many methodological issues that can set the future research agenda for PRD studies. In the remainder of this chapter, I will focus on the following issues: developmental stages, a multiple spatial perspective, the uniqueness of the PRD model, and interdisciplinary analyses.

Developmental stages

Using quantitative methods to analyze the data provided by the Guangdong Statistical Bureau, Shen *et al.* make a significant contribution by identifying local investment as the most important factor to explain the rapid economic growth of the PRD. Following this direction of analysis, one wonders whether it is also possible to identify the different stages of development in the PRD, the timing and characteristics of these stages, and the reasons for the transformation from one stage to another.

For example, several chapters in this volume hint that there may be four phases of development in the PRD: agricultural production, labour-intensive industry, capital-intensive industry, and globalization and high-tech industry. Borrowing W.W. Rostow's famous scheme and using the rate of development as the criterion, researchers can also identify the following phases in the PRD's development: stagnant economy, pre-conditions for take-off, take-off, sustained economic growth, and high consumption economy.

On the changing phases of development, many Western social scientists would like to know whether Hong Kong and the PRD have changed before, during, and after the 1997 transition, and whether the Asian financial crisis in the late 1990s had any significant impact on the region.

The PRD economy has developed very rapidly since 1978. Since the region has undergone more than two decades of development, now should be the time to examine the contour of PRD development, to trace the continuity and transformation, to identify the critical stages, and to understand the underlying causes for the success and failure of the PRD model in order to make better policies in the future.

A multiple spatial perspective

At first sight, the territorial boundary of the PRD seems to be quite clear. One can easily identify the PRD region on a map, and the chapters in this volume do focus on the PRD region.

However, studies of the PRD frequently need to go beyond the spatial boundary of the delta in order to explain the activities in the region. For example, factors at the national level need to be frequently referred to so as to explain the border issues. Hong Kong not only shares a border with the PRD, but it also shares a border with the central government in Beijing as well. Hong Kong must accept the central government's interpretation of the Basic Law on the right of abode issues, and it has to enlist the support of the central government before it can effectively control illegal migration and smuggling. Beijing's designation of Shenzhen as a pilot high-tech development area has certainly altered the path of development in the PRD. Beijing's policy toward Taiwan also has profound implications on Hong Kong's development. It seems that the more hostility that exists across the strait, the more autonomy the Hong Kong SAR government will have because Beijing will want to use Hong Kong as a show-case for national unification.

In addition, studies of the PRD need to refer to factors at the Asian regional level as well. The rise of Singapore as a regional financial center is often seen as a challenge to Hong Kong's global city status. The Taiwan government once encouraged its businessmen to invest in Southeast Asia rather than to invest in South China, a policy that could have important bearings for the development of Guangdong and Fujian. Needless to say, the Asian financial crisis in the late 1990s made Hong Kong realize how vulnerable it is if its economy is over-dependent on the financial sector.

Finally, studies of the PRD need to refer to factors at the global level. For example, China's entry into the WTO will further globalize the production, financial, and service sectors of China. Whether Hong Kong can maintain its status as the service center of China depends on whether it can compete with the transnational corporations in the Chinese markets.

The chapters in this volume discuss the above issues only in passing. Researchers still need to systematically study how global, Asian regional, and national factors interact with one another in the shaping of the development of the PRD.

The uniqueness of the PRD model

Several chapters in this volume adopt "a general approach" to the study of the PRD. In this approach, the PRD is described as a typical region in mainland China. This general approach assumes that what is found in the PRD can be readily generalizable to other cities, counties, and regions in mainland China. For example, Fan's chapter on labour market segmentation assumes that because Guangzhou has a segmented labour market, all other cities in mainland China also have such a feature. Similarly, Li and Siu's chapter on fertility assumes that because there are complex linkages between migration and fertility in Guangdong, such linkages can also be found in other provinces in mainland China as well.

On the other hand, there are several chapters which adopt "a unique approach" to the study of the PRD. In this approach, the PRD is regarded as a unique region. The PRD model of development is unique because the experience of PRD development cannot be generalized for the study of other regions. Instead of searching for similarities between the PRD and other regions in China, this unique approach sets out to determine in what ways the PRD is different from other Chinese regions. For example, the PRD is said to have a much faster rate of economic growth than other Chinese regions; its economic reforms are one-step ahead of others; its local governments have more freedom and autonomy than other local governments; it receives the largest percentage of non-local investments in China; its local residents have a strong lineage identity in the rural areas;

and its urban hierarchy includes an SAR and two Special Economic Zones, etc.

Adopting such an unique approach, researchers can examine how the PRD is different from other regions in China, and how such differences may help to account for the distinct pattern of development in the PRD.

Interdisciplinary analyses

Finally, the authors of this volume are mostly geographers and urban planners. Although their findings shed new light on the PRD region, their spatial analysis would be enriched if perspectives from other disciplines are also included.

For example, a political perspective of the PRD region would study how different modes of urban governance have promoted or diluted the conflict between temporary migrants and local residents in both the segmented labour markets and the urban enclaves. It would also be interesting to study how local government bodies have resolved or intensified the urban disputes arising from infrastructure construction (such as building a new football stadium, an electric plant, or a dam across a river) and urban planning (such as demolition of squatters, reclassification of the city-proper, or selling of land).

A social and cultural perspective could enhance our understanding of the PRD region as well. Sociologists and anthropologists have long emphasized the importance of family/kinship networks and social ties in the economic development of the PRD. Hong Kong industrialists prefer to invest in their native villages and districts in order to gain the social support of local governments and local residents. Also, social relations such as ethnicity and gender may play important roles in industrial relations and urban development in the PRD. A majority of the workforce in the PRD is made up of female migrants from other regions, and their gender and ethnic statuses should influence not only their fertility decisions but also their occupational aspirations, relations with their coworkers and bosses, work commitments, and attitudes toward local residents and government officials.

Thus, the complexity of the PRD region necessitates adopting an interdisciplinary approach to examine the interaction of spatial, political, social, and cultural factors in PRD development.

Conclusion

This volume has greatly advanced our understanding of the PRD. It shows that local investment — not non-local investment from Hong Kong — has played the most crucial role in the economic growth of the PRD. It argues that the long-term development of Hong Kong depends on closer integration with the PRD, and it thus criticizes the policy of a "super-fortress Hong Kong". It advocates that Hong Kong should treat Shenzhen as an equal partner — not as a satellite town — in order to promote high-tech development because Shenzhen is growing at a rate beyond the imagination of Hong Kong observers. It further advocates that the governments of the Hong Kong SAR and the local government in the PRD should take a more active role in promoting economic integration.

This volume also points to several problems of development in the PRD. Self-initiated migration has resulted in urban enclaves, segmented labour markets, and social polarization between temporary migrants and local residents. The overlapping and complicated structure of metropolitan administration in the PRD has led to massive duplications of infrastructure facilities and wastes of resources.

Thus far, political stability has been achieved by a "corporatist" mode of urban governance through which the state incorporates social forces into existing legal channels and successfully contains societal conflict to take place within the state arena rather than through street protests and public demonstrations.

To further advance the study of the PRD, it is proposed that future research projects should: identify the critical phases of PRD development and the transformation from one phase to another; adopt a multiple spatial perspective that examines the interactions among global, Asian regional, national, and local factors; explain how and why the PRD model is different from other regions in China; and adopt an interdisciplinary approach to trace the interactions of spatial, political, social, and cultural factors.